Y0-BUP-106

CHALLENGES OF ECONOMIC REFORM AND INDUSTRIAL GROWTH: CHINA'S WOOL WAR

Challenges of Economic Reform and Industrial Growth: China's Wool War

Edited by

Christopher Findlay

ALLEN & UNWIN

In association with
The Australia–Japan Research Centre
The Australian National University

First published in 1991
Allen & Unwin Pty Ltd
8 Napier Street, North Sydney, NSW 2059 Australia

National Library of Australia
Cataloguing-in-Publication entry:

Challenges of economic reform and industrial growth: China's wool war.

Bibliography.
Includes index.
ISBN 1 86373 203 9.
ISBN 1 86373 204 7 (pbk.).

1. Wool industry — China. I. Findlay, Christopher.

338.17631450951

Set by the Australia-Japan Research Centre, Australian National University, Canberra
Printed by SRM Production Services, Malaysia

Contents

Appendixes

Tables

Figures

Maps

Contributors

Kym Anderson	Department of Economics, University of Adelaide, Adelaide, Australia
Ray Byron	School of Business, Bond University, Gold Coast, Australia
James Crowley	University College Dublin, Belfield, Ireland
Christopher Findlay	Department of Economics, University of Adelaide, Adelaide, Australia
Melissa Gibbs	Department of Economics, University of Adelaide, Adelaide, Australia
Li Ze	Wool Textile Research Institute, Ministry of Textile Industry, Beijing, China
Liu Zheng	Rural Development Research Institute, Chinese Academy of Social Sciences, Beijing, China
Will Martin	National Centre for Development Studies, Australian National University, Canberra, Australia
David R. Thompson	National Centre for Development Studies, Australian National University, Canberra, Australia
Andrew Watson	Centre for Asian Studies, University of Adelaide, Adelaide, Australia

Preface

The results of research reported in this volume are the outcome of a long-term project involving a large team of researchers in Australia and China and reflect the productivity of joint work involving an active interchange of ideas and people.

The research was coordinated by the Australia-Japan Research Centre at the Australian National University as part of the Centre's continuing program of work on Australia's relations with Northeast Asia. It was supported by a three-year grant from the Wool Research and Development Fund, which was established by Australian wool-growers to support research and development in their industry.

The support of the Economic Research Advisory Committee of the Fund and its staff, especially Bob Richardson, John O'Connor and Chris Vlastuin, is acknowledged. The output of the project is reported in this volume, as well as in a companion volume on the international dimensions of the issues, entitled *New Silk Roads: East Asia and World Textile Markets*, edited by Kym Anderson and to be published by Cambridge. An earlier version of the overview chapter (Chapter 1) of this volume originally appeared as 'China and Australian wool' by Christopher Findlay and Andrew Watson in the *Pacific Economic Papers* series of the Australia-Japan Research Centre.

The research was conducted in cooperation with three Chinese research groups — the Wool Textile Research Institute in Beijing (attached to the Ministry of Textile Industry), the Institute of Rural Development Research, and the Institute of Industry Economics, attached to the Chinese Academy of Social Sciences.

As the research proceeded, drafts of papers were reported and reviewed at a series of workshops as well as at a final conference, all of which was supported financially and logistically by the Australia–Japan Research Centre. I wish to acknowledge the support of the Centre, its Research Committee, especially the Executive Director, Professor Peter Drysdale, and Centre staff, especially Martin O'Hare.

Other researchers were associated with the progress of the work. They included researchers from the Statistics Department in the Faculties at the Australian National University, and the Chinese Economy Research Unit at the University of Adelaide. Their own contribution, and, by implication, the infrastructural support of their host institutions, is evident in this volume.

The project benefited from the contributions of a large number of colleagues who happily participated in the workshops and in the final conference. Nicholas Lardy read the manuscript and provided an extremely constructive and detailed commentary.

The manuscript was wordprocessed by Debbie Beckman and prepared by Jan Holmes at the University of Adelaide. Gary Anson edited the work at the Australia–Japan Research Centre, where it was desktop published by Seán Knight. The index was prepared by Rosemary Thomson. Research assistance was provided by Steve Woodland, Stephen Chey, David Lawson and Joe Higuchi. The completion of this volume was made easier by the efficiency of the contributors. I have benefited from the assistance and advice of Peter Drysdale and Kym Anderson and the encouragement of, and contribution from, Andrew Watson.

Christopher Findlay
August 1990

1 Introduction and overview

CHRISTOPHER FINDLAY AND ANDREW WATSON

INTRODUCTION

The absence of a systematic program has been a distinctive feature of China's economic reform process. The Chinese did not set out to develop a step-by-step plan of reform to be phased in over a period of years. Instead, they adopted a number of strategic goals and in 1978 launched out on a path of piecemeal and pragmatic changes aimed at realising them. Essentially, the strategy adopted had four main aspects: a shift from economic growth expressed mainly through statistical targets towards emphasis on satisfying the consumption needs of the population; a change from extensive development based on new investment towards intensive development through greater efficiency; an acceptance of greater economic autonomy for producers, with a broader mix of methods of economic management and types of ownership; and the adoption of a much more open economy. The reforms adopted over the succeeding years have generally been consistent with these objectives, but they have not been implemented through a carefully planned series of stages. Overall, the process has been marked by different rates of reform across sectors, by occasional pauses and even retreats, and by problems generated by the interaction of the differing rates of reform. Enterprise managers, for example, have found that plan controls over their production or sales have disappeared at a faster rate than controls over their supply of inputs. Given the dual-price system and the continuing role of the central government in the supply of strategic materials and energy, the impact of the uneven pace of change on managers' behaviour has therefore been very complex.

Even if China had developed a careful program before launching its reforms, the process may not have been any easier. Three key issues are involved. These are the timing of each stage of reform, the linkages between each aspect of the reforms, and the influence of foreign reactions to the process. In a dynamic situation, many of the problems encountered in handling these issues cannot be anticipated. Furthermore, such problems not only include those related to economic shortages and imbalances but also those related to the political responses to the gains and losses created by the reforms and to the institutional changes that are necessary. Inevitably, there will be political debate and consequent adjustment to the program. The response of outsiders to China's economic changes are also beyond the control of China's reformers and yet play a significant role because of their feedback on the reforms as a whole. If outsiders do not help China participate in the world economy or if changes in international prices, perhaps as a result of China's entry into the market, work against China, it becomes more difficult for the reformers to proceed and to combat their opponents in the Chinese political system. Analysis of these issues of timing, linkages and foreign reaction and the way they interact will help explain the path of China's reforms to date and, thereby, also help anticipate the likely directions of the next steps of reform.

These issues have all been evident in the reforms in the rural sector, and one example which serves to highlight them is the raw wool industry and its user sectors in the textile industry. Because the raw wool industry is small and relatively less complex, it provides a particularly good case study. It has been deeply affected by the reform process, and the special features of sheep-raising and wool production provide a sharply delineated study of the way the reforms have worked. A further advantage of focusing on wool is that production is highly concentrated in the hinterland provinces. It involves consideration of the economic tensions in the relationship between the national minority areas engaged in wool production and the more developed coastal regions where the wool processing industry is concentrated. Finally, raw wool is an input into the light industry sector. It does not enter domestic consumption directly. This link enables us to examine the feedback between different sectors experiencing different rates of reform. This case study therefore offers the opportunity to analyse an aspect of the relationship between agricultural and industrial growth in a developing economy and to examine the determinants of a country's ability to supply the raw material requirements of industrialisation.

In the Chinese case, the problems of sequencing of reform led to a series of so-called commodity wars, among which the 'wool war' (Watson, Findlay and Du, 1989) is the focus of this volume. The demand for wool products had increased rapidly after 1978 as a result of rising domestic incomes and increasing exports of textiles and clothing. The increase in demand for finished products led to a rise in the demand for raw wool as well. But markets for finished products were relatively more free than markets for raw materials like wool, so finished product prices rose relative to raw wool prices. At the same time, meat markets were being reformed and meat prices were rising, leading to a higher slaughtering of sheep and a lower wool clip than otherwise. The demand for final products, the apparently low price of wool and the profitability of textile production supported decisions by regional governments, who had attained new levels of autonomy under the reforms in inter-governmental relations, to invest in wool textile production capacity. The consequent growth in demand for wool fibres put enormous pressure on the state commercial system which still dominated wool marketing. Attempts to by-pass that marketing system emerged, and a 'black', or secondary, market for wool developed.

The state in 1985 relaxed its control over the wool marketing system and the domestic price boomed. Access to the international market was restricted by the foreign exchange system which, although it permitted some considerable import growth, still limited imports below what otherwise would have been observed. Actors in the struggle for raw wool supplies included various types of enterprises, both within the plan and outside it, various levels of government and various kinds of merchants. In addition, wool-growers, who had attained higher levels of responsibility over their wool clip, became much more sensitive to relative prices.

The battles over the clip, however, did not solely take the form of competition and rising prices. Regional governments in pursuit of their perception of local interests raised barriers to the movement of wool within China. These barriers to inter-regional trade, as well as the processing of wool in local mills, which could be less efficient than the coastal specialists, raised the costs of this fibre war. In addition, while prices rose in response to market forces, there was little effort to reform other aspects of the wool marketing system so that the quality of the clip also fell during the period of the wool war (1985 to 1988). In summary, while price fluctuations are common for agricultural commodities like wool, in China these fluctuations were exaggerated by problems in the sequencing of reform. The subsequent struggle over raw wool supplies then reduced the efficiency with which wool was used in Chinese industry. Furthermore, the case of wool is but one example of a common problem. Similar stories can be told for other commodities such as silk, hemp and cotton.

After 1988 the wool war subsided. The cost of capital to the textile industry to purchase raw materials was by this time very high. The slower growth of the urban economy reduced consumer demand for wool products. By late 1989 textile producers were holding large stocks of finished products and commercial units were running down their holdings of wool goods. Some of the raw wool purchases during the previous years had been stockpiled, and these stocks were significant. Wool-growers were also holding substantial stocks of raw wool. These factors, and the higher world wool prices, contributed to China's withdrawal from import markets. Events such as a running down of wool stocks and a more rapid pace of reform and growth would force China back into the import market again. The fact that, in the face of lower wool prices and higher meat prices, herders in China are again reducing flock size also means that a shortage of domestic wool supplies is likely to arise again in coming years.

The sequence of events and the origins of the struggle are presented in more detail in the last chapter of this volume. The other chapters establish the context for the wool war. Chapter 2 reviews the factors determining domestic demand for wool products, which contributed to the struggle for raw material. Chapters 3 and 4 review the issues associated with the management of the foreign trade regime, and the reasons why, in the face of domestic supply problems, the cost of relying on world markets was also high. A broader set of issues related to self-sufficiency in agricultural products and the prospects for domestic fibre production are examined in Chapter 5. The recent events in the textile industry, especially the reasons for the rapid growth of the wool textile sector, are examined in Chapter 6, and some constraints on China's textile and clothing export marketing performance, and the implications of those constraints for the design of the reform agenda, are reviewed in Chapter 7. The focus in Chapter 8 is on the problems of the raw wool production sector. As noted, the pressures on the domestic marketing arrangements for raw wool and the policy responses are reviewed in the final chapter.

In this overview chapter, we outline most of the issues raised in the volume, and review some of them in detail. Our focus is on the problems of economic reform in China. The next section is therefore a short review of recent trends in the reform program and prospects for the 1990s.

PROSPECTS FOR REFORM

Some major successes were achieved in the first six years of the reforms after 1978, and standards of living increased rapidly. One important component of the reforms was the surge in China's participation in the world economy, including the textile, clothing and fibre trades.

However, from 1987 new problems emerged, especially in the agricultural sector. These problems included economic overheating, inflation and stagnating agricultural production, which were the result of both the rapid growth since 1978 and policy errors. The problems intensified the debate in China over the next steps in reform. By late 1988 there was a crisis in economic decision-making, and this contributed to the rising popular demand for political reform. The problems of economic reform had begun to merge with demands for transformation of the political system.

These policy debates of 1988 created severe divisions in the central leadership and contributed to the paralysis in decision-making. The subsequent slowdown in the reform process exacerbated the macroeconomic problems. Urban residents were already concerned about some of the implications of the reforms for their economic security and those concerns were heightened by the macroeconomic problems, especially inflation. Motivated by these economic interests, urban residents supported the campaigns for political reform and swelled the size of the demonstrations in April and May 1989. The Communist Party leadership then saw the situation moving out of its control, a perception which contributed to the severity of their reaction of early June of that year. The forceful repression of the demonstration did nothing to solve the underlying economic problems, and is likely to have exacerbated them.

Five main economic problems stand out:

1. the considerable overheating of the economy, which had already led to the adoption of a deflationary policy in late 1988;
2. the rising central government budget deficit, and the lack of central government control over national spending, due to the decentralisation of decision-making power;
3. the stronger identification of regional interests in China, which was leading to conflicts between levels of government over revenue, and exacerbating the conflict between state-owned and collectively-owned enterprises competing for raw materials;
4. a slump in agricultural production since 1984, related to the lack of progress in price reforms; and
5. the confusion of administrative power and economic functions, one result of which was considerable corruption.

The immediate response to these issues by the leadership was to slow down the growth of the economy and to recentralise some powers previously more widely distributed, with the general aim of controlling inflation and protecting state industries under central control. This reaction assisted the correction of the macroeconomic problems.

The experience of reform and growth since 1978 has, however, unleashed powerful new economic interest groups. The conflicting claims of these interest groups are becoming increasingly difficult to resolve within the present political structures. The growing inability of the structure dominated by the Party to resolve those conflicts led to the paralysis in decision-making and eventually the challenge to the regime as a whole. The net effect of this process is that there are now strong forces in society with a common interest in a reinstatement of economic policies orientated towards economic growth. Thus the Party-dominated regime is challenged by two issues: the efficient resolution of the conflicting claims emerging within the new sets of interest group relations in China, and the capacity to reinstate growth-oriented policies.

On both issues the present regime faces some contradictions. The Party structure is unwilling to entertain the claims of the new interest groups. At the same time, it faces great difficulties in reasserting macroeconomic controls and solving problems generated by reform, while still adopting policies that will promote growth. The contractionary macroeconomic policy is slowing down rural growth and undermining the development plans of regional and local governments, for example. The issue, then, is whether this regime can also re-create conditions for sustained high real growth rates, once it has brought the economy under control.

The political repression that has taken place to bolster the current leadership has undermined its longer run legitimacy and makes it less able to mobilise the initiative and creativity required to solve these economic problems. The Party's lack of credibility with its own members means that it now has great difficulty in formulating any coherent policies. Thus the tension between growth and control is likely to emerge again. Already in the Chinese press there are references to the need to accelerate economic reforms.

The prospects for reform are an important component of the outlook for the textile industry, and for its raw material suppliers. Income growth, and the demand for textiles and clothing, depend on the reforms. As noted above, the sequencing of the reforms will affect the adjustments to that income growth. The following two sections review the links between domestic income growth and demand for wool products and then examine a number of issues relating to the transmission of export market demand into the Chinese economy.

GROWTH IN DOMESTIC DEMAND FOR WOOL PRODUCTS

In the Chinese wool textile sector, an average of 15 per cent of output is exported and 85 per cent of output goes to the domestic market. The export orientation of individual subsectors is generally less than 10 per cent, and only in the knitted goods subsector is it very high, with around 50 per cent of output going to exports.

However, the export prospects of the sector are still important, for reasons discussed in the next section.

Turning now to domestic demand, income levels in China are still low, at around US$500 per head. The typical budget shares are about 50-60 per cent on food, 10-15 per cent on housing, 10-20 per cent on consumer materials and 10-15 per cent on clothing. The evidence presented by Byron in Chapter 2 is that the budget share of clothing is likely to remain at about 14 per cent for the next decade. As incomes rise, the absolute amount of expenditure on clothing will also rise but in proportion with the growth of income. Thus rapid rises in income, as occurred in the period after the reforms, can lead to equally rapid growth in domestic textile and clothing demand. If income doubles over the next decade, clothing consumption expenditure will also double. Information on consumption across income levels in China suggests a very rapid rise in fibre consumption as income rises, but, more importantly, wool fibre consumption is expected to rise faster than that of synthetic or cotton fibre. In other words, while the clothing share in spending may be constant, the share of wool in that clothing expenditure is likely to rise. Even within the wool budget, there will be a switch towards higher quality products. As a result, demand for wool in terms of volume may not rise as quickly as income.

Wool consumption in China is still at very low levels relative to cotton (about 1/3 of a kilo per head per year for wool, including blends, compared to 4 kilos per head per year for cotton and 2 kilos per head per year for synthetics) and there is scope for substitution out of cotton into wool but also into synthetics. The issue of fibre substitution in clothing consumption as income increases is important. The fibre preferences of local consumers will affect the pattern of specialisation of the Chinese textile industry. If China fails to improve the quality of its domestic wool production, there may well be increasing economic and political pressures to shift to other fibres in local textile production and this in turn is likely to influence the pattern of domestic consumption as well. This issue is discussed in more detail below.

There are some other points to make about the determinants of demand. There is significant variation in the structure of demand between the regions of China. In the countryside, for example, expenditure on clothing seems to be far less responsive to income growth. Second, an increase in the degree of urbanisation, also associated with overall economic growth, reinforces the demand for wool products in particular. Finally, demand varies between households in a systematic way, according to the stage of the family lifecycle and the employment status of the household head.[1]

ISSUES IN FOREIGN TRADE REFORMS

It was argued earlier that a combination of economic and political forces in mid-1990 led to the expectation of a continuation, in the medium term, of the economic reforms. In that case, a number of issues arise in the management of China's international economic relations. These are reviewed here under the headings of foreign exchange reform, protection in export markets, and the management of international business.

Foreign exchange reform

Of special interest are the implications of developments in the foreign exchange regime for the textile industry and the raw wool sector. Currently, the management of foreign exchange remains highly centralised. A movement to a more flexible regime is conditional on further market-oriented reforms in the rest of the economy. This is necessary, otherwise reform of the foreign exchange system alone could lead to an even less efficient allocation of resources and to new macroeconomic difficulties.

As argued above, the process of market-oriented reforms has slowed, so the date at which foreign exchange reform will occur may now be even further away. The significance of the foreign exchange system can be illustrated by asking the question, what if there was a foreign exchange reform?[2]

China is currently operating with a highly overvalued exchange rate. This is illustrated by the fact that, until recently, rates in secondary markets have been almost twice the official rates. Even since the devaluation of December 1989 there is evidence that the currency is still overvalued. The overvaluation of the official rate, and consequently the higher price of foreign currency in secondary markets, has increased the cost at the margin of importing raw wool. Some raw wool imports have been subsidised but these subsidies seem to apply only for infra-marginal purchases. In summary, the foreign exchange system has raised the domestic currency price of importables, such as raw wool.

The textile and clothing sectors are especially disadvantaged because they are not only raw material importers but also processed product exporters. They lose from the higher price of importables, such as raw wool, purchased using foreign exchange bought on the secondary market. They also lose from having their foreign currency earnings from exports converted at the overvalued official rate. They can retain some of these earnings, for sale in the secondary market if they wish, but that rate is certainly less than 100 per cent, especially given the important role of the state in controlling textile exports.

As a result of these arrangements, a devaluation of the official rate is expected to have the following effects. First, there would be an increase in the volume of clothing and textile exports; second, this would lead to increased demand for raw wool imports; and third, while the effects of devaluation on domestic prices of textiles and clothing are uncertain, there would be a positive overall effect of the reform on Chinese gross domestic product, which would contribute to a growth in domestic demand for textiles and clothing. The uncertainty about the domestic consumer price effects arises because the textile and clothing sector exports final products but imports intermediate products. Prices faced by Chinese consumers therefore change according to the higher returns from exporting, due to the devaluation of the official exchange rate, compared to the lower cost of imported intermediate products, whose prices fall as a result of the cheaper cost of foreign currency in secondary markets following devaluation.[3]

Protection in export markets

Would further increases in China's exports of labour-intensive manufactured products be possible without either depressing world prices of these products or triggering a protectionist responses in China's major markets? This question has been considered in detail by Anderson (1990a), who argues that there are grounds for a positive answer.

First, China's share of industrial countries' markets for these sorts of products is still very small, of the order of 1.6 per cent of total consumption of textiles, clothing and footwear, and less than 1 per cent of other light manufactures (in 1986). Second, as China enters the market, its competitors will be forced out to some extent, so the net addition to supplies by China's exports will be reduced. Furthermore, Anderson argues, China seems likely to be able to earn sufficient foreign exchange through exporting light manufactures to pay for all the other products, including fibres and other agricultural products, that it is likely to import.[4]

The conclusion is that there is scope for China to export larger volumes to world markets for these items. Nevertheless, the protective arrangements, particularly the Multifibre Arrangement (MFA), will have a significant effect on China's export performance (Anderson, 1990a, 1990b).

Some countries have achieved rapid export growth under the MFA, but there has been a substantial tightening of the provisions of the MFA in recent years. The evidence is that exporters, both newcomer suppliers like China and even established suppliers like others in Northeast Asia, will gain from the removal of the MFA (Trela and Whalley, 1988; Whalley, 1990). However, it is not only exporters of textiles and clothing who stand to gain from the removal of the MFA. If exports grow then fibre demand also grows. This means there is a common interest between raw material suppliers, some of whom are industrialised countries, and textile and clothing exporters like China, in the removal of the MFA. A coalition of interest between some industrialised and developing countries is potentially a significant force for removal of the MFA.

A major trade policy issue in this regard is that China is not a member of GATT. Membership would provide insurance against a protectionist response by the industrialised countries to an export surge by China. A condition of GATT membership, however, will be that the process of domestic reform continues in China. China can benefit from GATT membership in its export markets, but to do so it must accept constraints on its policy choices in domestic markets. Thus the interest in China in participating in world markets brings with it potential constraints on domestic policy choices, and because of the tightness of the foreign exchange constraints, it also brings with it some forces for a reinstatement of the reform program (Whalley, 1990).

At the same time, as argued by Anderson in Chapter 5, an outward-oriented policy in China implies a decline in China's self-sufficiency in various agricultural products, including natural fibres. China will be wary about that change because of the perception of vulnerability to events in world markets. The more signals that other countries can send to China to counter this perception, the more likely will be an outward orientation, and hence a more rapid reform of the domestic economy. The progress of the reforms and their contribution to growth thus depends a great deal on China's relations with trading partners, not only agricultural product suppliers, but also, for reasons explained in this section, countries which are importers of textiles and clothing.

The management of international business

China's export performance depends not only on the extent of protection in its major markets but also on the marketing system for China's products. The importance of this marketing system is demonstrated in Chapter 7 by Crowley, Findlay and Gibbs. They argue that the supply of marketing services to Chinese exporters is a determinant of their export performance.

Export marketing services need not be supplied from within China. They could be imported. Of course, Hong Kong is the traditional supplier of these services to China by re-exporting Chinese products, but in this context the MFA takes on added significance.

The MFA limits imports by country quota and imposes strict rules on 'country of origin'. The MFA therefore restricts not only trade in goods which are directly subject to its regulations, it also affects the supply of marketing services from Hong Kong. Because of the export quotas and the country of origin rules, China must either import less of these services from Hong Kong and therefore export less textiles and clothing than would otherwise be the case, or else seek other suppliers of those services who are not limited by MFA rules. It is not clear, however, that these other suppliers will be as competitive as Hong Kong from the Chinese point of view. In other words, China may have to pay a higher marketing margin to suppliers other than Hong Kong. These results highlight the economic significance of Hong Kong to China (see also Sung, 1988, and Chai, 1988), a feature which will be important in the evolution of the relationship between the two economies between now and 1997.

Second, marketing services could, in principle, be supplied from within China. If this is to be successful, a couple of other issues in the reform program will have to be tackled.

The problems of the management of China's exports are often discussed in terms of the 'air-lock' between Chinese enterprises and world market forces. It is argued that the institutional arrangements, particularly the role of the foreign trade corporations and the methods of pricing goods for export, has insulated enterprises involved in production from the world market-place. Thus it will also have isolated them from the incentive to deal with the marketing issues.

Attempts to improve the performance of the Chinese export sector will also focus attention on issues related to the transport and communications infrastructure in China. There are expected to be positive returns to investment in the transport and communications infrastructure.

However, over the time-scale relevant to assessing China's prospects in penetrating foreign markets, large-scale investments are not likely to be significant. More important in the shorter run will be the incentives of local exporters to work in and around the existing infrastructure. The demands for services that they create will also focus attention on the shortcomings of the infrastructure and make it more likely that efficient infrastructure investment programs will be pursued. According to this argument, the institutional arrangements for the management of international transactions, and the incentives that are created, will become increasingly important determinants of China's export performance, both now and in the longer run.

Clearly, these are important issues confronting China's textile exports. Such exports are one of China's major sources of foreign currency earnings, and statements during 1990 indicate that China's leaders expect them to grow and become even more profitable. In achieving that goal, issues of quality, fibre blends and composition and product differentiation have already been placed on the agenda by China's Ministry of Textiles. The role of synthetic fibres in the upgrading of the products of the textile industry is now under active discussion in the Chinese media.

THE TEXTILE INDUSTRY

The wool textile sector has grown much faster than the rest of the textile and clothing industry but it remains small relative to cotton (see Chapter 6). The prospects for the wool textile sector are positive, because it is relatively labour-intensive and therefore suits China's likely patterns of comparative advantage. However, different subsectors of the wool textile industry have grown at different rates. In particular, the knitting yarn sector and the woollen fabric sector (which use the 'woollen' rather than 'worsted' production process) have grown relatively quickly. The reason these woollen sectors have done so is because they have been highly profitable, and the most rapid growth in capacity occurred in the countryside. Since the woollen fabric-making process is relatively labour-intensive, it is attractive as a first step in the process of industrialisation.

Over recent years, however, there have been some problems in the wool textile sector of declining profitability and a falling capacity utilisation. The latter fell from an average of 94 per cent in 1978 to 77 per cent in 1987, and to even lower levels in the worsted subsector, down to 65 per cent. These problems were exacerbated by the slump in domestic demand as a result of macroeconomic policies adopted in 1989.

A source of the longer run problems has been the constraint on raw wool supplies. The degree of import penetration has risen to over 60 per cent and China's share of world imports of wool have also risen rapidly. But imports, as noted in the previous section, have been made more expensive by the foreign exchange system, so there had also to be other responses. First, there has been an increase in the use of chemical fibre in the wool textile industry, rising from 35 per cent of total fibres used in 1978 to 50 per cent in 1987. In 1987 total fibre consumption in the industry was about 360 mkg, of which about 180 mkg was (scoured) wool. The rapid penetration by chemical fibres into the wool textile industry is significant, and possibly a signal of a longer run trend, especially given the rapid growth of domestic production of chemical fibres, the growth in chemical fibre imports and the rise in their share of total fibre consumption by the textile industry as a whole to 30 per cent in 1989 (or about 2,000 mkg), up from 17 per cent in 1978. Recent plans by China's Ministry of Textiles call for a further increase to 32 per cent by 1992 and for increased use of blends in the wool textile sector.

Second, the composition of wool fibre consumption in the industry as a whole has shifted towards coarse wools. This reflects the more rapid growth of those subsectors, such as knitting yarn, which can more readily use local wool.

Third, domestic wool prices have been bid up, squeezing profits and putting pressure on the wool marketing system, as discussed in the description of the wool war earlier in the chapter.

Recent developments in parts of the wool textile sector have some influence on the overall outlook. The knitting yarn sector has had a number of advantages. It sells its produce direct to the public, especially the hand-knitting component of the yarn sector, so it has more access to 'free markets'. This permits it to sell it at higher prices and earn higher profits. Also, it has been able to use the coarser wool, which has been relatively more available in China. Furthermore, the system of hand-knitting yarn is the most forgiving in terms of technical faults. Producers of machine-knitting yarn can arrange connections to the knitted goods sector, which is relatively export-oriented and can provide the foreign exchange to buy the raw wool that is required.

The worsted sector, in comparison, is much more constrained by current policies. It consumes the highest quality wool, relatively scarce in China, especially such wool without faults. It is not surprising therefore that worsted capacity utilisation is relatively low.

This comparison yields a couple of implications in relation to the Chinese industry's search for raw material. First, the resolution of the raw material supply problem is not necessarily solely in the hands of the textile industry. As argued above, there are major problems in the management in the foreign trade regime that inhibit imports and force the industry into other forms of adjustment. Second, the overall process of economic reform is clearly critical for the development of this industry. Third, a further implication is that raw material suppliers who export to China are highly likely to try innovative marketing activities, such as various forms of counter-trading, which can provide raw material, without triggering the administrative processes of the foreign exchange system.

These problems of supply of natural fibres, combined with the issues involved in reliance on world markets, discussed above, raise questions about the path of fibre specialisation in the textile industry. As already noted, the use of chemical fibres to produce blended-fibre products in the wool sector has increased rapidly. In this regard, comparison with the rest of Northeast Asia is of interest. The typical trend has been to substitute out of the production of natural fibre textiles into synthetics (Anderson and Park, 1989). For example, in Korea and Taiwan, synthetic fibres now account for 70 to 80 per cent of total fibre consumption. But in China these shares are reversed. Japan has followed a different path to the rest of Northeast Asia: natural fibre still accounts for about half of total fibres consumed. One question for China is whether it will follow a path of rapid substitution into synthetics like Korea or Taiwan or the slower path like Japan. It is argued below that this choice will depend in part on raw material supply arrangements, both within China and from the world market.

Another issue is whether there are technological constraints facing the industry. In terms of potential competitiveness at factory level, because of the labour intensity of this industry and the availability of labour in China, the answer might be no. On the other hand, the international developments in the textile industry and the marketing of textile products suggest an increasing degree of sophistication (see Chapter 7). These developments will require high-speed communication between purchasers of products and their producers, a tighter product specification, more flexible production systems driven from consumer demands, and feedback down to the specification of the fibres required. The question is, can the Chinese industry keep up with this development, especially on the export side? The evidence so far is that it cannot. The industry's domestic market orientation, and its sales of large lots of standard items, implies a different managerial approach to that required in the export market. This problem opens up scope for some complementary trade not only in technology but also in marketing. Obviously, Hong Kong plays a critical role, not just in relocating activities in which it is losing competitiveness, but also in providing those marketing and technology services, as discussed in the preceding section.

In summary, the Chinese textile industry faces some important choices in relation to its fibre specialisation. Will it continue to consume natural fibres, which implies a rising share in world imports of these fibres, or will it pursue a policy of greater self-sufficiency, by consuming more locally produced synthetic fibres? As argued by Anderson in Chapter 5, China can expect to see a decline in self-sufficiency in both wool and cotton fibres. As a result, China's reliance on world markets for these fibres will increase. But the execution of that strategy is constrained by domestic policy problems, especially in relation to the foreign exchange management system, the reform of which is also related to China's perceptions about the reliability of world markets. Furthermore, as argued in the next section, the quality of the domestic supplies of raw wool, and the manner in which they are presented to the textile industry, raises the costs of using domestic wool. Incentives presently exist therefore to switch out of natural fibres, even though this may also involve the transformation of the textile industry into more capital-intensive methods of production (see Chapter 6). The fibre mix in the textile industry will depend in the longer run on the progress of the economic reforms, and on developments in the Chinese raw wool sector.

THE RAW WOOL SECTOR

The main issue in the raw wool sector is the low productivity of wool producers in China. Many factors influence the willingness of Chinese growers to produce wool, such as the prices of alternative activities and prices of inputs into wool production. The main interest in this section is, however, the set of factors which might lead to wholesale shifts in wool production, whatever the price.

The Chinese sheep flock is large, of the order of 100 million wool sheep. The bulk of this flock is located in the pastoral areas in the northern parts of the country. In those areas, summers are warm and moist and winters are very cold and dry. As a result there is a high degree of seasonal variation in the growth of feed. The conditions are summarised by the cliche that the sheep are 'full in summer, fat in autumn, thin in winter and dead in spring'. The region is also prone to drought and excessive snow.

There are three main topics for consideration. First is the management of the pasture, second is the investment in the rural infrastructure, and third is the influence of management practices compared to genetics as contributors to wool output (see Chapter 8).

With respect to pasture land, there has been a significant degradation of its quality in China. A number of factors have contributed—for example, cultivation, especially in the period of the Cultural Revolution's deforestation and the cutting of trees for fuel, and more recently the introduction of the household responsibility system. Furthermore, the long-term trend has been to increase total flock size while decreasing pasture area.

In recent years, sheep numbers in China have been rising, in part associated with the movement to household responsibility. But that system also involved the removal of some of the controls over the use of pasture land. There is now concern that sheep numbers are so high as to be unsustainable. Numbers may have to drop below what otherwise might have been possible in order to regenerate pastureland. The implication is that increasing wool output will not be possible by increasing the number of sheep. Increased wool output will only be possible by improving the productivity of animals; that is, the conversion of grass into clean wool.

It might be expected that, having identified the problem, there would be some policy reaction—for example, enclosing grasslands and the introduction of a new management regime. Indeed, it is the case that regional governments in China now have a greater degree of autonomy than previously and they could act to correct these problems. But the incentives to do so are not present. The incentives offered by the tax-revenue arrangements and the fiscal arrangements with the central government are to invest outside agriculture (Watson, 1989). Local governments do not earn high returns from maintaining the rural infrastructure. The same point applies to veterinary services, and extension services, all of which are also made more expensive because of the fragmentation of ownership of livestock.

Of course, the idea was that higher prices paid to growers as a result of the reforms and the profits would be used to fund local investment. There were two problems with this plan. First, it did not work for publicly consumed assets. Peasants, for example, would rather build a house than invest in a common pastureland. Second, increased income from higher wool and meat prices have gone directly to households and there are few mechanisms to aggregate this capital for investment.

The third major topic is the question of the contribution of genetics compared to management as contributors to wool output. There are two broad sets of possibilities. One is that new genetic material could be introduced into household flocks, and two, that management practices could be improved.

Considering the management issue first, there are some challenges to face. The first set is those of the incentives created by the marketing system, especially the lack of variation of prices in relation to quality and the fact that wool prices are fixed by raw weight. Clearly, this marketing problem contributes to the low clean yields (that is, the proportion of clean wool in the total weight of a fleece) in China, which are about 30-40 per cent and falling. A related issue is that the local wool textile industry faces many problems and higher costs as a result of the impurities in the clip and the inconsistencies in the fibre quality within their purchases. The difficulties in handling the domestic clip, exaggerated because of inefficiencies in the raw wool marketing system, reduce the incentive to specialise in wool products. Another problem is the fragmentation of the sheep flocks. The flocks are now mixed by age, type and sex, leading to dilution of quality as well as making the diffusion of any new technology very difficult.

With respect to genetic material, given the loss of quality in sheep flocks, output could also be improved by a better selection of growers and improved even faster by the injection of higher quality animals. But the question is where would those animals come from? China already has many pure breeds comprised of big sheep, which have high greasy yields and high survival rates. The problem is to spread them around. The introduction of foreign genetic material may not have any immediate effects at the farm level. The introduction of Merino genes, for example, has led to increased greasy yield and staple length. On the negative side, it has also led to a drop in body weight and the lambing rate. To achieve an animal suitable for local conditions, it is necessary to develop a new Chinese breed.

The experience of cross-breeding with Merino sheep highlights an important issue — the definition of the goals of the breeding program, especially the importance of meat. The demand for lamb meat is increasing with income, especially in the cities. On the other hand, the sheep farmers, mainly minority groups and particularly Muslims, prefer to eat mutton from local mature sheep. So there will have to be a balance between wool and meat, and between young and mature meat production (which influences the age structure of the flock). The balance selected will reflect the different political interests of the local minority and the more urban Han Chinese population. If there is a preference to be given to meat, especially lamb, production in the breeding program, then it is likely to lower wool output volume and increase the degree of coarseness in the clip. It is to be expected that any Chinese breeding program will have to take account of meat output needs.

In principle, another method of increasing wool production is to extend the area of land devoted to running sheep. This means increasing sheep numbers in southern China. Current discussion of this issue in China stresses the use of hilly grasslands not fit for farming. It talks of a large usable area with high herbage yields, and suitable for all-year grazing. Nevertheless, the development of this resource will require the introduction of appropriate breeds and management technologies for higher rainfall areas. Much research needs to be done. Climate and opportunity costs may make the shift uneconomic. For example, wool and meat production will have to compete with grain production for land and with other animal husbandry activities for feedgrain. The opportunity cost of grain, as perceived by Chinese policymakers, will become increasingly high as China's food self-sufficiency falls as expected (see Chapter 5). Already, Chinese policymakers are deeply concerned by the decline in per capita grain availability since 1984 and the likelihood of that worsening over the next ten years. Although there has been some discussion in China of restoring to pasture the land diverted to grain production since 1949, the major problems of ensuring sufficient grain supplies for the growing population (already acknowledged to be likely to exceed the target of 1,200 million by the year 2000) mean that there is minimal potential for shifting large amounts of land and labour to wool production.

In the pastoral areas, the main alternative to wool production is for its joint product, namely meat. High meat prices, as in 1989, encourage higher slaughter rates, a younger average age of flock and lower wool production. The growth in meat production which is expected to follow the high prices is likely to lead to falling wool production and a running down of stocks. The prospects are therefore for the emergence of another 'wool war' in a couple of years.

In summary, in the immediate future, any major gains in productivity are likely to come from changes in wool growing management practices. Those changes in turn are dependent on the progress of the price and marketing reforms in China. But the binding constraint is the supply of pastureland. There is limited scope to increase the land area or its grass yield. In the long term, a stable level of wool production in China will depend on achieving a balance between pasture productivity and sheep numbers, improved management practices and genetic change to increase yield per animal. Even so, genetic inputs or management changes are likely to have very small effects relative to the growing gap between the demand for raw wool in China, driven by rising demand for wool products in domestic and export markets, and its supply.

2 Demand for wool products in China

RAY BYRON

INTRODUCTION AND OVERVIEW

Demand for fibres in China can be expected to grow not only because of export growth but also because of the growth in domestic demand for finished clothing and textiles. The growth in demand will be stimulated by the increase in income levels. In addition, higher incomes and the increasing degree of urbanisation could lead to a change in the preferences for different types of textiles (such as suiting material) and different types of fibres. These issues are reviewed in this chapter by summarising the results of studies of data on clothing consumption over time in some Northeast Asian economies and by reporting results of analysis of household expenditure data. The first section of this chapter contains a review of the factors affecting domestic demand and a summary of the results of the statistical analysis. The detail of the statistical analysis and the methodology employed are presented in the second section of the chapter.

Income levels in China are still relatively low, about US$500 a year (Perkins, 1988). The typical household budget in China is mostly spent on food (50 to 60 per cent), followed by consumer durables and other goods (15 to 20 per cent) and then clothing and housing (each 10 to 15 per cent). The very high food share is not surprising given the low income levels. This comparison of income levels and expenditure patterns is qualified by the debate over estimates of per capita income in China. The World Bank estimates are even lower than Perkins's, and other estimates are far higher (Summers and Heston, 1988).

To obtain some idea of the likely magnitude of the growth in China's domestic demand for clothing, especially for natural fibres, it is helpful to look at the growth of fibre use in other Northeast Asian economies.

In the early 1960s, Korea used about the same quantities of natural fibres per capita as China did in the early 1970s, that is, around 2.7 kg of cotton and 0.11 kg of wool each year (see Table 2.1 and Figure 2.1). By the early 1970s, Korea's use of cotton and wool had risen to 4 kg and 0.3 kg respectively, which again was almost the same as China's consumption in the early 1980s. This physical evidence of consumption levels suggests the gap between Chinese and Korean

Table 2.1 Per capita consumption of raw cotton and wool, China, other East Asian economies and the world, 1961-86 (kg)

	China	Korea	Taiwan	Hong Kong	Japan	World
Cotton						
1961-64	1.80	2.70	6.30	37.10	8.9	3.4
1965-69	3.20	3.00	7.00	41.00	8.5	3.5
1970-74	2.90	4.20	9.70	38.20	8.3	3.4
1975-79	3.10	8.30	12.70	46.40	7.5	3.2
1980-84	3.90	10.40	14.60	30.40	7.6	3.3
1986	4.90	9.70	16.60	24.00	6.6	3.5
Wool						
1961-64	0.15	0.11	0.10	0.10	2.9	0.69
1965-69	0.13	0.27	0.54	0.49	3.3	0.67
1970-74	0.15	0.48	1.06	0.76	3.34	0.62
1975-79	0.15	0.90	1.09	0.63	2.37	0.55
1980-84	0.24	1.25	1.21	0.34	1.79	0.52
1986	0.33	1.42	1.27	0.43	1.85	0.54

Note: Production plus imports minus exports of raw fibre.
Source: International Economic Data Bank, *Agricultural Tapes* (based on USDA and FAO sources), Australian National University, Canberra, 1989.

incomes is smaller than indicated by official statistics (for example, World Bank, 1989).

Over the period from the early 1970s, China thus appears to be on the same track that Korea was following only a decade earlier. But what has happened to Korea's fibre consumption since the early 1970s? The answer to this question may give us some guide to the pattern of China's future fibre use. Since the early 1970s, Korea's use of cotton has doubled again while wool use has grown even faster, trebling since the early 1970s. This suggests that if rapid growth can be maintained in China, fibre consumption, especially wool fibre consumption, will grow rapidly.

What other evidence is there concerning the likely growth in fibre consumption? One set of data sources are surveys of household expenditure patterns. For example, one study of the composition of household expenditure in China in 1982 found the following variations in spending across income levels, reported in Table 2.2 in index number form.

The wool products category, which includes both yarn and clothing, increased rapidly as the income level per head of household rose. This suggests that as more households move into higher income classes over time, the demand for wool products can be expected to increase relatively rapidly.

In summary, both the experience of the rest of Northeast Asia and evidence on household expenditure in China suggests a strong relationship between

Figure 2.1 Per capita consumption of raw cotton and wool, China and other East Asian economies, 1961–86

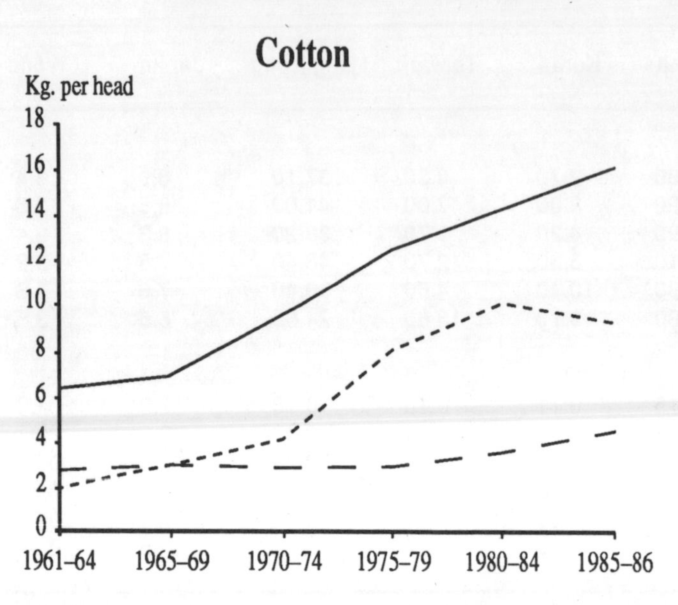

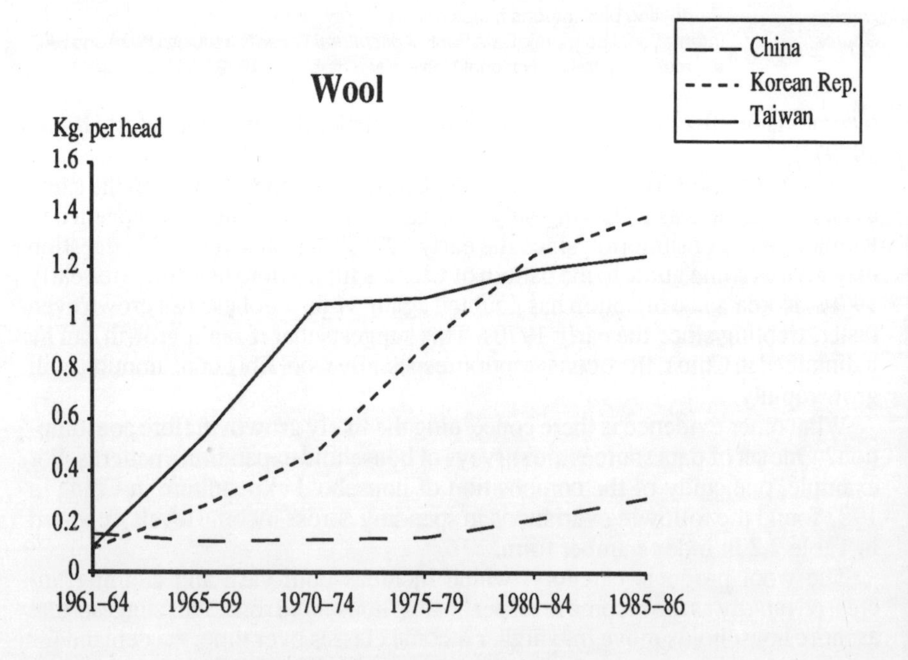

Note: Production plus imports minus exports of raw fibre.
Source: International Economic Data Bank, *Agricultural Tapes* (based on USDA and FAO sources), Australian National University, Canberra, 1989.

income and clothing consumption, especially wool consumption. The data presented so far are simply descriptive and have no statistical significance.

Table 2.2 Variation across household income levels in clothing consumption by fibre type (index numbers)

	Low	Mid	High
Income level index	100	200	300
Cotton clothing	100	200	200
Synthetics	100	266	366
Wool products	100	250	500

Source: Lu Xuezeng, Yang Shengming and He Juhuang (1985).

However, statistical results reported in the next section suggest that the clothing share of expenditure remains roughly constant while the wool share in clothing expenditure could rise. There are of course a number of factors which will cause some variation around these trends.

One is that there is considerable variation in clothing consumption between regions of China. Data quoted so far are derived mainly from surveys of urban households, and the expenditure patterns of peasant households are different. This is partly the result of lower incomes on average in the countryside and partly the different lifestyles. The data suggest that consumption of clothing as a whole is much less responsive to income growth than in the urban areas. Consumption categories which grow more rapidly than income in the countryside are housing and consumer durables. Partly, this also reflects the provision of subsidised housing to urban residents. This result also implies that as China becomes more industrialised, and therefore more urbanised, in the process of economic growth, there could be a shift in preferences towards wool products.

Household expenditure data also suggest there is considerable variation between households in their spending decisions. In other words, while on average there is a significant relationship between income and clothing consumption, especially wool products, there is a lot of variation between households. In particular, the demographic characteristics of the household, such as age, are important influences on decision-making.

There is a connection between the type of employment and the demand for wool products. For example, households where some family members are in more senior management positions are more likely to have suits. In some cases, firms in China will give a subsidy to their employees to make the purchase. Another category of households likely to be observed buying a suit are those where a son in the family has recently been married. Generally, there will be a correlation between age of the head of the household and the decision to buy a suit. Whether or not the item purchased is a wool product could also depend on household income levels.

As argued so far, growth in wool demand in China can be largely explained by the rapid economic growth of the last ten years. Economic reforms have not only raised the general standard of living for most Chinese people, they have also allowed more freedom of choice for consumers and manufacturers, along with more responsibility.

Increased demand for wool is also affected by other social factors. Some of the factors which have been studied elsewhere (Chey and Lau, 1989) include population growth, growth in information channels, increased contacts with foreigners and with the outside world, changes in social class, and general consumer awareness of the choice in quality of goods available for purchase (in other words, fibre consciousness).

China's population growth is well known. To control this, the government has enforced with a fair degree of success the single child policy. But following the introduction of this policy, family spending on consumer items for children, including toys and clothes, has greatly increased.

Information channels have multiplied in China and advertising is increasingly powerful. Advertising of Woolmark has been tested and shown to be very influential in the cities where it has been introduced in the last four years. Survey results indicate that relatively more educated people and those with more disposable income are likely to be more discriminating in the purchase of textile products. Access to Foreign Exchange Certificated currency is one measure of privilege and can be shown to be influential in wool purchases (Chey and Lau, 1989).

Chinese consumers are becoming more aware of fibre content when purchasing fabric or garments. Stores usually advertise fibre content, and prices are based on this factor. Wool has benefited from this situation up to now, but reforms in pricing policy and in the retail system may not continue to support the growth of this fibre consciousness.

With the opening up of China to the outside world, the Chinese urban population has also changed its attitude to foreign styles of clothing and other aspects of lifestyle. Western-style clothing is more suited to the use of wool than traditional Chinese clothing, particularly for the tailored suit, and to the general use of knitwear, so this change in aspirations and lifestyle has also favoured wool.

The balance of this chapter reviews existing statistical analyses of household expenditure survey data and reports results of new statistical work.

HOUSEHOLD EXPENDITURE STUDIES

Review of existing studies

Projecting the demand for any commodity in China is made difficult by the absence of data and the rapidity of institutional change over the last decade. Household expenditure surveys, for example, exist on a provincial and national basis but the data is published at a highly aggregated level. The scarcity of data makes it difficult to obtain reliable estimates of the income, price and demographic effects which are critical for forecasting. Lack of data also restricted the study reported below to the demand for clothing rather than the demand for specific fibres.

There have been previous studies projecting consumer demand in China, all of which adopted the same methodology—the Extended Linear Expenditure System (ELES). Previous results, especially in relation to the demand for food and clothing, are discussed in this section. Some expenditure relations are then estimated based on seven years of household expenditure data cross-classified into six or seven income groups for urban areas and into 28 provinces in the case of rural areas. The urban study has the advantage of some additional demographic information on household composition which is lacking in the rural data. The rural data show extraordinary differences in consumption characteristics between provinces.

He Juhuang (1985), in a paper based on household surveys for Beijing and Hubei Province in 1981 and 1982 (14-15 observations), used the ELES, which has the advantage that it is possible to estimate the effect of price on demand, even though data on prices may be absent. This is a property of all systems derivable from a strictly additive utility function as the income compensated price effects are proportional to the income effects. The disadvantages, such as the exclusion of complements and inferior goods and the restrictions imposed on substitution possibilities by the nature of the Klein-Rubin utility function, are well known. In addition, looked at from a long-term perspective, the linearity of the Engel curves make this function rather unattractive.

He's estimates of marginal budget shares, subsistence expenditure levels, own price and income elasticities for Beijing and Hubei are given in Table 2.3. The author does not provide estimates of the standard errors and, given the fact that the estimates are from grouped data and his estimation technique does not allow for grouping (see Maddala, 1977: 268-74, for instance), the results should be treated with caution. In such a situation one can anticipate that the r^2 terms will be biased towards unity and the standard errors will be biased downwards. Nevertheless, despite these cautionary remarks, it is worth recording the marginal budget shares for subsequent comparison. The results for urban and rural areas were basically similar.

Table 2.3 Marginal budget shares for Beijing and Hubei, 1982

	Food	Clothing	Fuel	Other goods	Services
Beijing	0.54	0.18	0.01	0.21	0.05
Hubei	0.49	0.16	0.04	0.15	0.02

Source: He Juhuang (1985).

He Juhuang makes a number of useful observations about difficulties in applying conventional demand models to the Chinese data. Expenditure on services is low and constant, and spending on food does not appear to decline with income as might be anticipated. These features could be attributed to free items in the service sector, such as medical care, education and housing as well as the lack of availability of consumer goods, and even foodstuffs. The effect of rationing in the sample period could make projections based on such data quite hazardous. The marginal budget shares ($\partial p_i q_i / \partial m$)for food, clothing and other goods were 0.54, 0.15 and 0.18, with income elasticities of 0.90, 1.26 and 1.39 respectively. The income elasticity of demand for clothing is quite high and could even be understated because of the presence of rationing. On the other hand, these data refer only to purchased items of clothing and exclude, for example, rural self-consumption of fibres for household production and consumption.

The second study, by Van der Gaag (1984), utilises per capita consumption data published in the *State Statistical Yearbook* for China in 1981 and 1982. The data are for rural households in 28 provinces, excluding Tibet. Simple linear and share-log Engel curves are fitted, despite the differences in consumption patterns and climatic influences in the northern and southern provinces. The data were pooled, a time dummy variable added and the square of income (or the log of income) was included as an explanatory variable. The squared terms were generally successful in the share-log equation but, with the exception of fuel, not in the levels equation.

Income elasticities, calculated at the mean, are presented for the linear forms of both equations and are given in Table 2.4 for later comparison. The coefficients were not particularly well determined; nevertheless, the income elasticities for housing are much higher than observed previously whilst the income elasticities of demand for food and clothing were lower.

Table 2.4 Income elasticities, pooled data for 1981-82

	Linear	Share-log form
Food	0.76	0.80
Clothing	0.67	0.72
Fuel	0.49	0.78
Housing	2.69	2.37
Daily articles	1.33	1.31
Services	0.73	0.60

Source: Van der Gaag (1984).

Engel curves—theoretical considerations

Engel curves have the longest history of any functional form in econometrics. Despite this, the area can still provoke intense debate. Early Engel curve researchers were extremely concerned about shape and form of Engel curves (within the constraints of linear estimation), but it is probably true to say that the current generation has placed a greater emphasis on consistency with utility theory (in other words, being a special case of a well behaved and perhaps integrable demand system), satisfying the budget constraint and consistency with aggregation theory over commodities and/or individuals. The simple Linear Expenditure System (LES) is

$$(1) \quad v_i = \alpha_i + \beta_i m + u_i$$

where $v_i = p_i q_i$ is the share of total expenditure (m) devoted to good i. The slope coefficient $\beta_i = \partial v_i / \partial m$, that is, the 'marginal budget share'.

A different form of Engel curve is the Working–Leser model, which is a semi-log form with expenditure shares as the dependent variable (Deaton and Muellbauer, 1980: 75-80). The Working–Leser model has the Engel form

$$(2) \quad w_i = \delta_i + \gamma_i \log(m) + u_i$$

When m = 1, log(m) = 0 and $d_i = w_i$ is the average budget share. The slope coefficient

$$(3) \quad \gamma_i = \frac{\partial w_i}{\partial \log(m)} = \frac{\partial v_i}{\partial m} - \frac{v_i}{m} = \beta_i - w_i$$

In other words, $\beta_i = w_i + \gamma_i$, and γ_i is the difference between the average and the marginal budget shares. Expressed as a relationship between shares and the log of income, the Working–Leser model has a reasonable variety of shapes.

The empirical work presented in support of the Working–Leser model is almost invariably based on grouped data and, as such, is somewhat suspect with r^2's biased towards unity. When disaggregated survey data are used, the author (amongst others) has noticed a tendency for more complicated forms, such as quadratic versions of the Working model, to be supported.

These two simple forms of Engel curves are used below. More sophisticated forms (such as logistic based models) yield no perceptible benefits with the small sample of grouped data available. To compare the different systems, rather than use non-nested hypothesis testing in such a small sample situation, the error sum of squares of each system are compared in the others' metric.

A RE-ESTIMATION OF ENGEL CURVES FOR URBAN AREAS

Despite misgivings about LES, a logical place to start this empirical exercise is that very point, by using weighted least squares to compensate for the fact that the data are grouped. The problem with grouped data is that one is estimating a regression equation based on grouped averages

$$(4) \quad \overline{y} = \alpha + \beta \overline{X} + \overline{u}$$

Then, if Var $(u_i) = \sigma^2$, Var $(\overline{u_i}) = \sigma^2/n_i$. The introduced heteroscedasticity is counteracted by multiplying each term by $\sqrt{n_i}$ and the equation to be estimated becomes

$$(5) \qquad \sqrt{n_i}\,\bar{y}_i = \alpha\sqrt{n_i} + \beta\sqrt{n_i}\,\bar{X}_i$$

Note, in passing, the well known result of Prais and Aitchison (1954) that the estimator from grouped data is always less efficient than from ungrouped data and that the r^2 from grouped data will generally be much higher than from a regression on ungrouped data. Haitovsky (1972) cites results in which the r^2 for ungrouped data with 1,218 observations jumps from 0.03 to 0.91 after grouping.[1]

With these warnings, a set of cross-section studies for urban areas in China can now be examined. The data are given in Table 2.5 while some results are given in Tables 2.6–2.9. Figures 2.2 and 2.3 summarise the data from Table 2.5. They show a declining share of food expenditure as a proportion of income as income rises. The picture for the share of clothing expenditure in income is less clear and both relationships require further statistical analysis, as below.

The following presents Engel curves based on the LES and the Working model. These functional forms have the advantage of being able to represent a variety of responses to changes in income, expecially when the income term is introduced quadratically. The LES is applied to this data with three commodities based on food, clothing, and other goods and services. The equations in Tables 2.6 to 2.8 are the results from weighted least squares regressions of each good's expenditure (per capita) on total expenditure (per capita and the square of total expenditure per capita and demographic variables. The equations are of the form

$$(6) \qquad p_i q_i = v_i = \alpha_i + \beta_i m + u_i$$

$$(7) \qquad p_i q_i = v_i = \alpha_i + \beta_i m + \delta_i m^2 + u_i$$

$$(8) \qquad w_i = \alpha_i + \beta_i \ln m + u_i$$

$$(9) \qquad w_i = \alpha_i + \beta_i \ln m + \delta_i (\ln m)^2 + u_i$$

where m is income and w is the expenditure share.

For comparison, the error sum of squares based on expenditure levels sse(u) and shares sse(v) are given. The residuals are transformed to these measures before computing the sum of squares. Based on the two sse measures, there is little to choose between the LES and the Working model; they tend to yield similar error sum of square terms. However, taking t-statistics on the individual coefficients into account, and the Breusch–Pagan LM test for heteroscedasticity, three different specifications of the expenditure system are preferred here. In the case of food, the preferred form is quadratic in income with total household members as the demographic variable. The linear form is appropriate for clothing, but with the number of wage earners in the household as an additional explanatory variable.

Table 2.5 Household income and expenditure for workers in cities[a]

	1982	1983	1984	1985	1986	1987
Number of households	83	55	209	1 714	3 113	87
	332	269	1 315	1 714	3 113	307
	2 312	1 841	4 861	3 429	6 225	803
	4 095	4 218	2 834	3 429	6 225	1 734
	1 281	1 488	1 601	3 429	6 225	2 559
	917	1 109	1 680	1 714	3 113	2 450
				1 714	3 112	1 776
Av. number of persons per household	5.88	5.49	5.77	4.47	4.59	2.61
	5.22	51.3	4.84	4.19	4.48	2.88
	4.62	4.52	4.19	3.99	4.97	3.32
	4.02	3.99	3.80	3.80	3.82	3.45
	3.80	3.75	3.71	3.62	3.63	3.6
	3.34	3.55	3.48	3.48	3.47	3.75
				3.24	3.19	3.88
Av. number of employees per household	1.71	1.78	1.88	1.82	1.73	0.62
	1.93	1.96	2.05	2.03	2.06	0.93
	2.12	2.12	2.22	2.12	2.07	1.50
	2.37	2.33	2.37	2.22	2.13	1.78
	2.78	2.59	2.49	2.30	2.22	1.92
	2.79	2.82	2.72	2.43	2.26	2.03
				2.41	2.25	2.15
Av. number of dependants	3.44	3.08	3.07	2.46	2.69	4.18
	2.70	2.62	2.36	2.06	2.18	3.09
	2.18	2.13	1.89	1.88	1.97	2.21
	1.70	1.71	1.60	1.71	1.79	1.94
	1.37	1.45	1.49	1.57	1.63	1.88
	1.20	1.26	1.28	1.43	1.54	1.85
				1.34	1.42	1.81
Annual income[b]	231.24	242.16	248.88	437.70	446.40	398.76
	296.40	304.68	410.40	546.72	570.36	536.04
	401.40	407.64	558.72	632.88	663.24	627.84
	538.44	548.16	712.44	737.28	772.92	751.08
	697.08	706.32	839.16	861.96	905.04	859.02
	877.68	895.20	1 081.68	1 012.32	1 054.56	936.36
				1 276.20	1 347.12	1 062.84
Annual total of living expenditure[b]	227.64	230.16	263.52	455.64	469.80	385.20
	269.04	282.00	361.56	551.28	573.96	503.04
	362.04	371.76	479.76	626.88	650.99	580.08
	473.52	487.08	601.80	724.20	747.36	666.12
	598.92	614.04	705.12	830.28	855.24	756.60
	724.08	759.60	887.04	963.24	987.48	836.52
				1 162.92	1 262.88	914.16

Table 2.5 (continued)

	1982	1983	1984	1985	1986	1987
Annual expenditure on commodities[bc]	209.28	210.24	242.16	418.92	431.76	343.76
	244.44	259.08	332.00	505.92	527.64	466.32
	330.84	340.68	441.36	576.12	597.96	530.88
	434.52	447.00	553.32	668.52	686.88	610.56
	552.36	564.48	648.12	766.44	789.48	691.08
	681.12	696.60	814.44	892.08	911.04	764.53
Annual expenditure on food[bc]	143.64	142.08	161.88	278.88	275.52	264.12
	171.96	178.92	220.92	319.56	323.64	328.08
	220.32	227.16	285.12	352.44	356.28	345.12
	277.44	290.64	348.60	392.76	395.76	385.08
	340.20	335.80	398.99	430.80	434.64	422.64
	409.68	432.60	486.00	473.88	482.28	450.86
				546.12	572.16	486.48
Annual expenditure on clothing[bc]	29.28	34.08	39.36	60.00	58.68	31.08
	33.24	38.28	54.24	81.36	75.36	50.88
	48.72	52.80	75.60	95.76	91.80	66.48
	68.64	71.52	94.92	113.52	108.24	88.92
	93.00	92.16	110.04	130.56	124.44	101.76
	115.92	107.52	133.08	152.64	141.24	114.00
				178.68	166.32	125.76
Annual non-commodity expenditure	18.36	19.92	21.36	36.72	38.04	31.44
	24.60	22.92	29.40	45.24	46.32	36.72
	31.20	31.08	34.80	50.76	53.04	49.08
	39.00	40.08	48.48	55.68	60.48	55.56
	46.56	49.56	57.00	63.84	65.76	65.52
	60.96	63.00	72.60	71.16	76.44	72.00
				85.80	93.48	79.56

Notes: a Data referring to the same urban area are in the same cell position in each category.
b In per capita terms.
c Other goods and services (see Table 2.8) is defined as annual expenditure on commodities minus expenditure on food and clothing.

Source: *State Statistical Yearbook*, various issues.

Finally, for other goods and services, the results support a quadratic system with both number of household members and number of wage earners included. Note, as would be anticipated, that the usual properties of adding up systems are observed. The preferred equations are highlighted in bold. The Breusch and Pagan (1979) heteroscedasticity test would have us all reject the null hypothesis, in the case of the clothing equation, but as four other tests pointed to acceptance and the sample size was low for asymptotic inference, the evidence of heteroscedasticity was inconclusive.

Figure 2.2 **Annual income vs. food as a proportion of annual income**

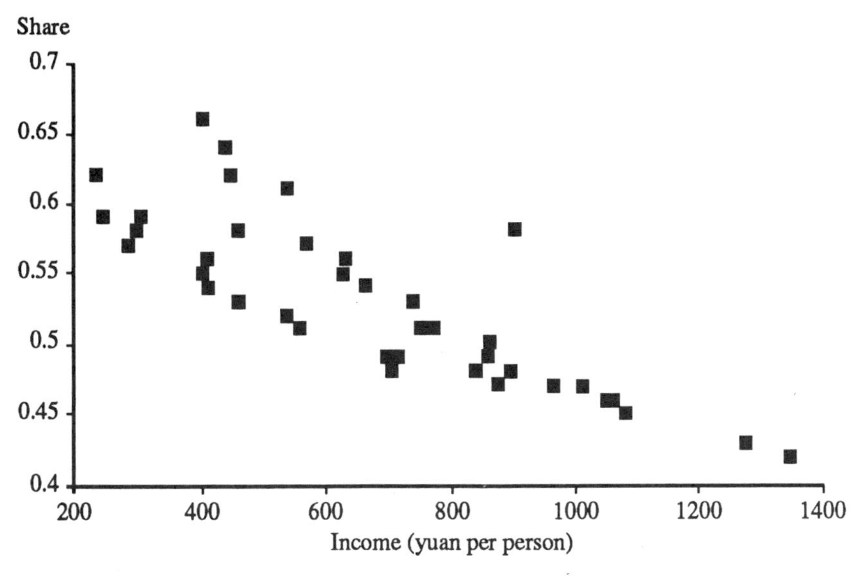

Source: Table 2.5.

Figure 2.3 **Annual income vs. clothing as a proportion of annual income**

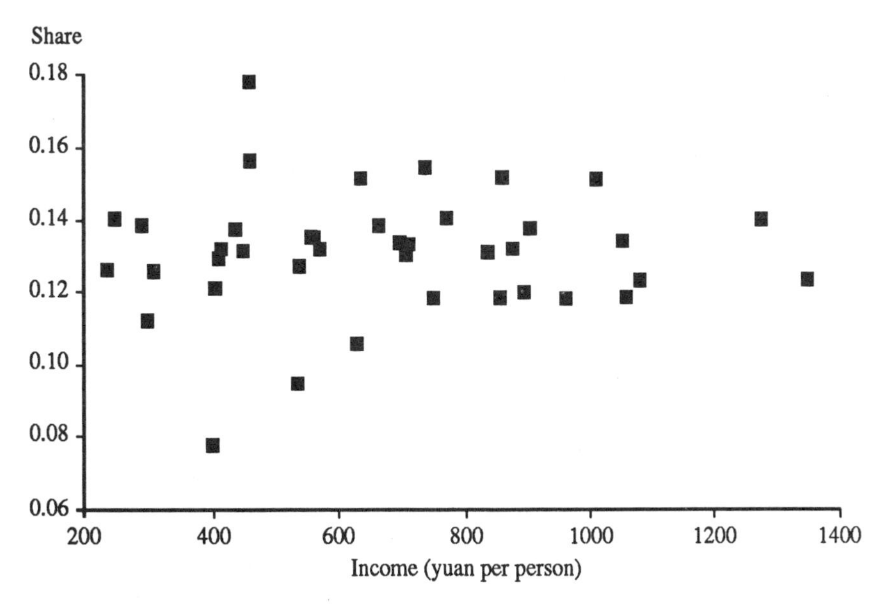

Source: Table 2.5.

Table 2.6 Expenditure on food, weighted least squares estimates and t-ratios

				Expenditure on food			
exp	0.403	0.382	0.402	0.381	**0.624**	0.652	0.625
	38.0	26.8	35.8	26.4	**16.6**	16.8	16.4
exp²					**-0.15^{-3}**	-0.16^{-3}	-0.15^{-3}
					6.7	6.5	6.6
no hh		-13.7		-13.8	**-9.75**		-10.14
		2.2		2.2	**2.3**		2.4
no earn			3.54	-0.753		0.105	-2.89
			0.4	0.08		0.02	0.4
const	94.0	163.7	87.0	165.8	**62.1**	8.22	69.56
	12.2	5.1	4.2	4.0	**2.4**	.48	2.2
r²	0.975	0.978	0.974	0.978	**0.990**	0.998	0.990
sse(u)	7560	6626	7532	6625	**2918**	3380	2900
sse(v)	0.1325	0.0660	0.1309	0.0663	**0.0073**	0.0178	0.0063
het	2.9	6.1	4.6	6.3	**6.3**	11.6[a]	6.6
				Share of expenditure on food			
lnexp	-0.122	-0.140	-0.121	-0.139	0.566	0.584	0.571
	15.8	13.7	15.0	13.6	3.0	2.8	3.0
lnexp²					-0.054	-0.054	-0.055
					3.8	3.4	3.8
no hh		-0.016		-0.017	-0.016		-0.017
		2.5		2.6	2.9		3.1
no earn			-0.0028	-0.0078		-0.0035	-0.0085
			0.28	0.8		0.4	1.0
const	0.872	0.891	0.872	0.893	0.922	0.904	0.925
r²	0.872	0.891	0.872	0.893	0.922	0.904	0.925
sse(u)	7274	5612	7170	5461	3565	4746	3364
sse(v)	0.0084	0.0071	0.0084	0.0070	0.0051	0.0063	0.0049
het	7.7[a]	8.5[a]	17.1[a]	7.1	19.2[a]	18.3[a]	8.2

Note: a Rejects the null hypothesis in the Breusch-Pagan heteroscedasticity test.

Table 2.7 Expenditure on clothing, weighted least squares estimates and t-ratios

				Expenditure on clothing			
exp	0.143	0.137	**0.137**	0.136	0.190	0.184	0.184
	27.0	18.6	**31.8**	23.0	6.8	8.9	8.3
exp^2					-0.3^{-4}	-0.3^{-4}	-0.3^{-4}
					1.9	2.3	2.2
no hh		-3.54		-1.02	-2.67		-2.9
		1.1		23.1	0.9		0.12
no earn			**18.0**	17.7		17.4	17.3
			4.9	4.6		4.9	4.8
const	1.03	19.0	**-34.6**	-28.8	-3.3	-49.5	-47.7
	.27	1.1	**4.4**	1.7	0.17	5.1	2.7
r^2	0.951	0.953	**0.971**	0.971	0.958	0.975	0.975
sse(u)	1845	1783	**1109**	1103	1604	959	958
sse(v)	0.01.2	0.0136	**0.0029**	0.0032	0.1057	0.0072	0.0075
het	0.20	20.7[a]	**11.6[a]**	11.0[a]	23.6[a]	8.3	8.4
				Share of expenditure on clothing			
lnexp	-0.0036	-0.5^{-3}	-0.0027	-0.0024	0.210	0.191	0.191
	0.7	0.06	0.66	0.4	1.3	1.6	1.6
$lnexp^2$					-0.016	-0.015	-0.015
					1.3	1.7	1.6
no hh		0.0036		0.3^{-3}	-0.004		0.2^{-3}
		0.7		0.07	0.3		0.06
no earn			0.029	0.03		0.028	0.028
			5.6	5.3		5.7	5.5
const	0.120	0.162	0.099	0.095	-0.512	-0.526	-0.528
r^2	0.012	0.028	0.471	0.473	0.074	0.512	0.514
sse(u)	2815	3028	1174	1168	2972	1155	1149
sse(v)	0.040	0.0039	0.0021	0.0021	0.0038	0.0019	0.0019
het	0.99	25.6[a]	2.5	5.5	26.2[a]	9.0	10.3

Note: a Rejects the null hypothesis in the Breusch-Pagan heteroscedasticity test.

Table 2.8 Expenditure on other goods and services, weighted least squares estimates and t-ratios

	Expenditure on other goods and services						
exp	0.452	0.481	0.460	0.482	0.185	0.163	**0.190**
	36.4	29.5	37.0	30.2	4.7	4.2	**5.1**
exp²					-0.20^{-3}	-0.19^{-3}	**-0.18^{-3}**
					7.7	7.8	**8.1**
no hh		17.3		14.9	12.5		**10.5**
		2.5		2.1	2.9		**2.5**
no earn			-21.5	-16.9		-17.5	**-14.4**
			2.0	1.6		2.7	**2.3**
const	-95.0	-183.1	-52.4	-137.4	-59.4	41.1	**-22.4**
	10.6	5.0	2.3	3.0	2.1	2.3	**0.7**
r²	0.972	0.977	0.975	0.978	0.991	0.991	**0.993**
sse(u)	10205	8717	9150	8096	3227	3296	**2781**
sse(v)	0.1413	0.0739	0.1227	0.0583	0.0144	0.0154	**0.0125**
het	1.1	0.24	3.9	5.8	2.7	15.1[a]	**2.2**
	Share of expenditure on other goods and services						
lnexp	0.118	0.141	0.124	0.142	-0.773	-0.774	-0.761
	13.1	11.9	13.9	12.4	3.7	3.6	3.9
lnexp²					0.070	0.069	0.070
					4.4	4.1	4.6
no hh		0.020		0.017	0.020		0.017
		2.6		2.3	3.2		2.9
no earn			-0.025	-0.020		-0.025	-0.020
			2.3	2.0		2.7	2.3
const	-0.466	0.689	0.446	0.641	2.25	2.44	2.26
	7.9	6.95	7.9	6.4	3.3	3.5	3.6
r²	0.825	0.851	0.845	0.865	0.904	0.897	0.915
sse(u)	7232	6348	6710	5397	3834	3542	2650
sse(v)	0.0117	0.0098	0.0102	0.0088	0.0020	0.0068	0.0054
het	10.4[a]	8.5	17.6[a]	11.3[a]	5.2	19.7[a]	2.7

Note: a Rejects the null hypothesis in the Breusch-Pagan heteroscedasticity test.

Table 2.9 Seemingly unrelated regression weighted least squares estimates and t-ratios, expenditure equations

	Food	Clothing	Other goods and services
exp	0.651	0.138	0.212
	18.1	31.6	5.9
exp^2	-0.17^{-3}		-0.17^{-3}
	7.8		7.8
no hh	-9.91		9.9
	2.5		2.5
no earn		17.35	-17.35
		4.9	0.81
const	53.89	-33.26	-21.23
	2.2	4.4	0.81
r^2	0.998	0.996	0.996

The preferred specification for each equation was then adopted and combined using seemingly unrelated regression estimation. The results are presented in Table 2.9. Combined estimation made very little difference to the results.

The average and marginal budget shares for the three groups of commodities, based on the expenditure equations, are given in Table 2.10, and point to a constancy of clothing in terms of its share of total expenditure (and an income elasticity of unity) — a result borne out by the non-significance of the quadratic terms in the equations estimated. Projection then is relatively easy — based on these results, it is possible to predict a continued budget share for clothing, up to the end of the century, in the 0.135–0.145 range. This is also evident, with the benefit of hindsight, in Figure 2.3.

Table 2.10 Average and marginal budget shares and income elasticities (calculated at the mean)

	Food	Clothing	Other
Average	0.571	0.141	0.288
Marginal	0.440	0.138	0.413
Elasticity	0.797	0.968	1.357

ENGEL CURVES FOR CHINA

The data for this section were also drawn from the *State Statistical Yearbook* for China. The tables there tabulate living expenditure per capita by peasant households in 28 provinces over the years 1981–87. The expenditure groups are

Table 2.11 Average budget shares, marginal budget shares and t-values based on 1981-87 data

	Food	Clothing	Other		Food	Clothing	Other		Food	Clothing	Other
Beijing				***Anhui***				***Yunnan***			
ABS	0.513	0.114	0.373	ABS	0.613	0.100	0.672	ABS	0.099	0.229	0.288
MBS	0.432	0.074	0.495	MBS	0.514	0.060	0.692	MBS	0.048	0.261	0.426
t-value	15.663	10.816	16.708	t-value	17.507	7.781	15.967	t-value	2.800	8.569	0.124
r2	0.980	0.959	0.982	r2	0.984	0.924	0.981	r2	0.611	0.936	0.979
Tianjin				***Fujian***				***Shaanxi***			
ABS	0.505	0.122	0.373	ABS	0.647	0.081	0.608	ABS	0.117	0.275	0.272
MBS	0.507	0.058	0.435	MBS	0.621	0.031	0.503	MBS	0.096	0.401	0.348
t-value	24.329	7.551	23.338	t-value	27.359	5.296	24.880	t-value	7.879	31.176	0.128
r2	0.992	0.919	0.991	r2	0.993	0.849	0.992	r2	0.925	0.995	0.985
Hebei				***Jiangxi***				***Gansu***			
ABS	0.526	0.123	0.352	ABS	0.629	0.093	0.649	ABS	0.126	0.225	0.278
MBS	0.464	0.076	0.460	MBS	0.561	0.051	0.474	MBS	0.109	0.417	0.388
t-value	27.996	15.736	24.832	t-value	22.323	7.782	43.126	t-value	7.001	34.607	0.026
r2	0.994	0.980	0.992	r2	0.990	0.924	0.997	r2	0.907	0.996	0.978
Shanxi				***Shandong***				***Qinghai***			
ABS	0.584	0.144	0.272	ABS	0.518	0.124	0.662	ABS	0.143	0.195	0.358
MBS	0.487	0.112	0.401	MBS	0.518	0.070	0.545	MBS	0.153	0.302	0.412
t-value	18.826	8.151	15.859	t-value	31.267	20.309	21.816	t-value	7.389	12.583	0.168
r2	0.986	0.930	0.981	r2	0.995	0.988	0.990	r2	0.916	0.969	0.992
Inner Mongolia				***Henan***				***Ningxia***			
ABS	0.067	0.116	0.277	ABS	0.571	0.125	0.612	ABS	0.131	0.257	0.304
MBS	0.449	0.036	0.515	MBS	0.551	0.062	0.545	MBS	0.086	0.369	0.387
t-value	3.274	2.301	3.412	t-value	24.936	8.783	42.041	t-value	5.470	24.880	0.207
r2	0.682	0.514	0.700	r2	0.992	0.939	0.997	r2	0.857	0.992	0.989
Liaoning				***Hubei***				***Xinjiang***			
ABS	0.541	0.120	0.339	ABS	0.659	0.111	0.603	ABS	0.185	0.213	0.230
MBS	0.485	0.084	0.430	MBS	0.843	0.053	0.559	MBS	0.121	0.320	0.104
t-value	27.269	8.300	36.422	t-value	4.649	5.401	28.144	t-value	13.463	16.249	0.598
r2	0.993	0.932	0.996	r2	0.812	0.854	0.994	r2	0.973	0.981	0.067
Jilin				***Hunan***				***Guizhou***			
ABS	0.606	0.115	0.278	ABS	0.647	0.352	0.254	ABS	0.701	0.101	0.198
MBS	0.485	0.084	0.430	MBS	0.843	0.053	0.559	MBS	0.121	0.320	0.104
t-value	12.847	3.505	14.970	t-value	17.257	9.041	12.330	t-value	49.669	3.747	14.602
r2	0.97	10.711	0.978	r2	0.983	0.998	0.968	r2	0.998	0.737	0.977
Heilongjiang				***Guangdong***				***Zhejiang***			
ABS	0.582	0.135	0.282	ABS	0.616	0.060	0.324	ABS	0.549	0.100	0.352
MBS	0.576	0.048	0.376	MBS	0.616	0.016	0.368	MBS	0.474	0.050	0.476
t-value	23.891	4.384	14.669	t-value	39.998	1.833	29.454	t-value	41.715	7.143	50.041
r2	0.991	0.794	0.977	r2	0.997	0.402	0.994	r2	0.997	0.911	0.998
Shanghai				***Guangxi***				***Sichuan***			
ABS	0.479	0.090	0.432	ABS	0.663	0.078	0.259	ABS	0.661	0.102	0.238
MBS	0.432	0.074	0.494	MBS	0.528	0.021	0.396	MBS	0.602	0.055	0.343
t-value	34.334	9.836	24.301	t-value	29.294	1.778	0.396	t-value	39.152	3.901	23.312
r2	0.992	0.051	0.992	r2	0.994	0.387	0.985	r2	0.997	0.753	0.991
Jiangsu											
ABS	0.543	0.099	0.358								
MBS	0.442	0.048	0.510								
t-value	39.269	6.360	31.318								
r2	0.997	0.890	0.995								

food, clothing and other goods. Data on sample sizes and average number of household members were unavailable so the results below, based on seven observations, are exploratory only. The results are based on estimation of the expenditure equation:

$$v_i = \alpha + \beta\ m + u_i$$

The average and marginal budget shares for each province are listed in Table 2.11, as is the t-value on the MBS coefficient and the r^2 for the equation. What is of most interest are the differences between provinces in the average and marginal budget shares and the occasional large differences within a province between the average and the marginal budget shares. The results suggest one should not fit models to China as a whole—the regions have very different consumption characteristics.

In terms of the original data, the average budget share devoted to clothing is highest in the north and the inland regions; the average budget share on housing is highest in the northeast and the coastal provinces; and the average budget share devoted to food is highest in the central and western regions—presumably an indicator of a lower standard of living.

THE DEMAND FOR FIBRE TYPES

There is some information in the published results of the household surveys which enable relationships between purchases of items by fibre type and household income levels to be identified. It is not clear from the information given in the *State Statistical Yearbook* but the data apppears to have been collected in the same surveys of urban households reported in the third section of this chapter. Four years of data were available from 1982 to 1985, each year typically having six income classifications. The upshot of a number of regression equations was the conclusion that the income elasticities for wool cloth, woollen fabric clothes and knitted wool items are high, and in fact it is fairly safe to conclude they are greater than one. Cotton items appear to have quite low income elasticities, synthetics somewhat higher, and silks and satins rank with wool in elasticity terms. However, these results could be conservative in the light of the presence of rationing.

CONCLUSIONS

The statistical work reported here was hamstrung by the data available and had to be confined to an Engle curve analysis of clothing demand. The more sophisticated forms of Engel curves proved no more superior to simple functional forms in explaining the data. The demand for clothing appears to be linear over a wide range of income with marginal budget shares which are quite high, expecially in the urban areas. This is significant given the population drift that has and is still occurring. A significant, and unusual, determinant of the demand for clothing was the number of wage earners in a household and not the number of household members. In this respect also, the clothing category differed from the other two groups examined—food and other goods. The final section, presented in summary form, was based on a few scraps of information on the annual purchases of durable items extracted from the household expenditure survey and revealed remarkably high income elasticities of demand for wool items, much higher than for cotton or synthetics. One can only hope that more disaggregated and detailed household data will become available in the not too distant future so that a more comprehensive and conclusive study can be carried out.

3 Analysing the effects of China's foreign exchange system on the market for wool

WILL MARTIN

INTRODUCTION

The policy reforms of the past decade have greatly increased the importance of foreign trade and the foreign exchange system in the Chinese economy. Over this period, China has also become a much more important participant in the world market for wool and wool products, particularly as an importer of wool. Between 1980 and 1986 China's imports of virgin wool grew from 2.5 per cent of world imports to over 10 per cent (see Chapter 5).

China's foreign trade system, and its foreign exchange system in particular, have been identified as having a potentially major influence on wool imports (Angel, Simmons and Coote, 1988). The foreign trade system has profound implications for the entire economy, exerting an influence on the overall size of the economy and hence consumption levels and living standards. In addition, it affects relative prices between activities and hence the pattern of production and consumption. China's foreign trade system has undergone vast changes since 1978 and there is clearly a great deal of interest in further reform to facilitate economic growth and development (see Zhang, 1987; Dai Luzhang, 1987; Ho Lok Sang, 1986, for example). Lin (1988) concludes that it is widely recognised within China that the current foreign exchange system is untenable and that reform is essential for modernisation. An understanding of the implications of the current system is essential if the effects of future reforms are to be predicted.

The approach adopted in this chapter takes into account the fact that the foreign exchange system has ramifications throughout the economy, and these cannot be ignored in assessing its effects on a particular market. Further, it takes into account the fact that both central planning and market mechanisms coexist in the post-reform economy (Byrd, 1987, 1989; Sicular, 1988a).

Given the dual nature of China's 'planned commodity economy', a central issue is the extent to which prices affect the allocation of resources. In a fully centralised system where export and import levels are determined solely by planning decisions, pricing instruments such as the exchange rate and tariffs fulfill only an accounting function. By contrast, in a decentralised market system, these variables play a central role in the determination of exports and imports. Thus the first issue addressed in this chapter is the relative importance of plan and market in the relevant areas of China's economy.

The nature of the foreign exchange and foreign trade systems, and their implications for the economy, and for the fibre, textile and apparel sectors in particular, are discussed in the third section. The effects of the foreign exchange system on the economy as a whole are evaluated in the fourth section. The impacts on the fibre, textile and apparel sector are then examined in the fifth section. The results of this analysis suggest that foreign exchange reform will involve a complex set of effects on the economy. While some conclusions can be drawn using a simple qualitative framework, the complex nature of the interactions between markets makes it impossible to determine the overall effects of reform on the wool market. The first detailed attempt to evaluate this effect is reported in the next chapter in this volume, where a general equilibrium model is developed and used to analyse the complete effects of exchange rate policy changes.

PLAN AND MARKET IN THE CHINESE ECONOMY

A major thrust of the reforms in China's economy since 1978 has been to decentralise economic decision-making away from central planning and towards provincial governments and enterprises. This process has involved increased use of markets as a means of allocating resources, although planning by central and provincial authorities remains very important (Naughton, 1985; Wong, 1985), particularly in determining the distribution of gains and losses.

The extent to which production decision-making has been decentralised depends upon the sector of the economy. It appears, for instance, that decision-making has been substantially decentralised in the agricultural sector (Sicular, 1988a; Lardy, 1983a). Progress to date has generally been viewed as having been somewhat less in the industrial sector (Naughton, 1985; Wong, 1985; Chinese Economic System Reform Research Institute, 1987), although recent work suggests that the reforms have made free-market prices very important in influencing industrial sector decisions at the margin (Byrd, 1987; Wu Jinglian and Zhao Renwei, 1987). The foreign trade system has also been substantially decentralised, with successive phases of reform involving increased devolution of responsibility for individual decisions and the development of improved control mechanisms (Shan, 1989).

An important feature of Chinese economic policy since 1978 has been the introduction of two-tier (or multi-tier) pricing. Under this system, part of the total production takes place under a planning-type system in which specific quantities of production inputs and output are supplied at specified official prices, and part of the consumption requirements are met by allocations or rations at state determined prices. Most importantly, however, enterprises are frequently able to produce additional output at market, rather than official, prices, and consumers are able to buy or sell additional amounts at market prices.

As Byrd (1987), Wu Jinglian and Zhao Renwei (1987) and Sicular (1988a) have argued, in principle, only the prices prevailing at the margin have any effect on output levels under a system of legal secondary markets such as those developed in China, unless the planning system imposes constraints on particular outcomes. In fact, the existence of these legal markets greatly simplifies the analysis relative to the case of illegal parallel markets considered by Roemer (1986). This point can be readily illustrated for an individual market such as that depicted in Figure 3.1, where suppliers are contracted to supply a quantity q_o at official price p_o. In this figure, we assume, for simplicity, that consumers do not face any corresponding fixed price, although the argument can readily be generalised to incorporate fixed consumer prices associated with rationing or allocation systems.

Figure 3.1 Effects of two-tier pricing in an individual market

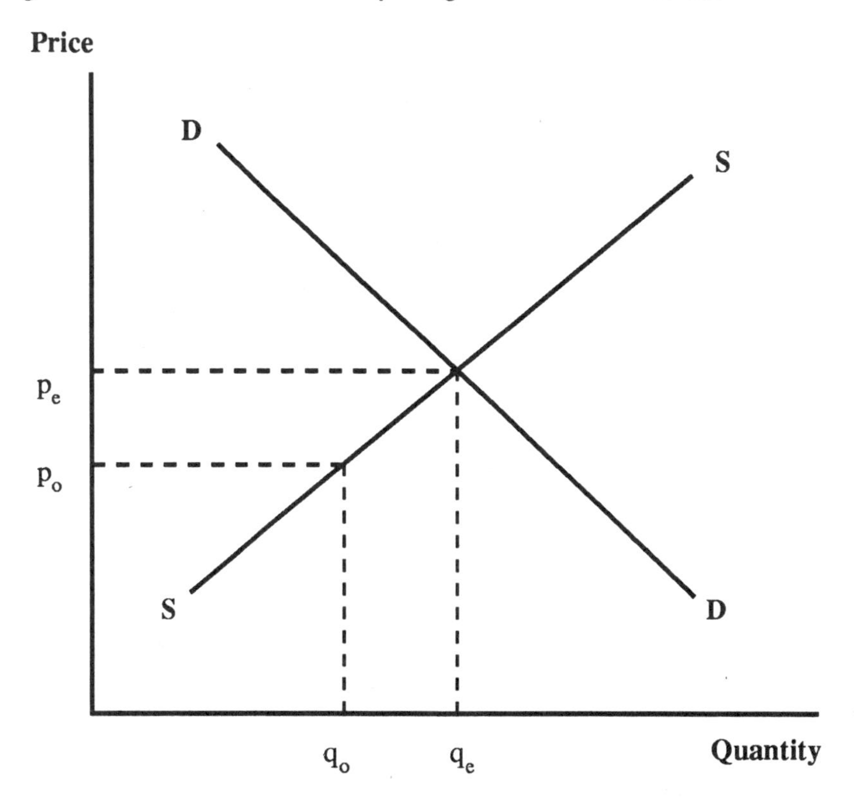

As long as q_o, and the state contract price p_o, do not affect the price obtained for output at the margin, they will not affect the position of the short-run supply curve SS in Figure 3.1. Thus the total level of output and the secondary market price p_e will be shown by the same intersection of SS and DD that would represent the price and quantity outcomes in the absence of intervention.

While the existence of a quota delivery amount q_o, or a corresponding contract to deliver a fixed quantity of revenue, has no impact on resource allocation, output or free-market prices in this market, it does generate a revenue transfer. In the case depicted in Figure 3.1, the transfer from the supplying enterprises is shown by the area $(p_e - p_o)\, q_o$. As Figure 3.1 shows, such transfers are perfectly non-distorting taxes whose only effect on resource allocation results from their effects on the distribution of income and hence demand. In practice, however, pursuit of these economic rents is likely to lead to significant rent-seeking costs. Further, the potentially large profits available to those able to reallocate products from the official to the free market create powerful incentives for illicit reclassification of commodities.

The extent to which two-tier pricing systems apply, and the institutional form of the two-tier pricing system, varies substantially between the agricultural, industrial and foreign trade sectors of the Chinese economy. While some scholars, such as Sicular (1988a) and Byrd (1987), emphasise the allocative function of markets in the post-reform Chinese economy, others point to a continuing role for planning decisions (such as Perkins, 1988: 615). Accordingly, each of the major sectors of the Chinese economy is examined in turn in the next three sections of this chapter.

The agricultural sector

The process of economic reform has had its greatest impact in the agricultural sector. The introduction of the household responsibility system has transferred much of the direct responsibility for decision-making to the enterprise level. While the planning authorities have imposed delivery quotas, at least until recently, the proportion of output marketed outside these quotas appears to have increased for most commodities, although Sicular (1988a: 287) reports that only 18 per cent of total agricultural production was sold at market prices in 1984.

The planning system has essentially involved a requirement for producers to deliver specified quantities of output to the state at set prices which, in the case of food, are generally above the net returns from the sale of that food. As Sicular (1988a) demonstrates, such a system will undoubtedly have powerful redistributive effects, but will not influence the allocation of resources, as long as the market prices relevant to marginal output decisions are not distorted and the resulting redistribution of income does not greatly affect demand patterns.

The planning system definitely inhibits the response of producers to price signals in a number of ways. These include a tendency for the 'negotiated' prices for over-quota production to be related to within-quota prices, uncertainty about the link between current production and future quota delivery or contract levels, and central allocation of some inputs such as fertilisers. While these factors could be expected to reduce the response of producers to changes in prices, they would not be expected to eliminate price responsiveness. Nevertheless, it seems clear from the available information on agricultural policy in China (for example, Sicular, 1985, 1987, 1988a; Lardy, 1983a), and from the observed response of

Chinese agriculture to major price changes in the early 1980s, that the Chinese agricultural sector now responds significantly to prices.

Table 3.1 Output, state purchases and purchase prices for cotton, 1970-87

	Output (1 000t)	State purchases (1 000t)	Cotton exports (1 000t)	Cotton imports (1 000t)	Apparent mill consumption[a] (1 000t)	Cotton purchase prices (yuan/100kg)	Cotton[b] above-quota price premium
1970	2 277	2 030	52	81	2 059 (99)	204	
1971	2 105	1 913	51	122	1 984 (96)	204	
1972	1 958	1 788	56	196	1 928 (93)	210	
1973	2 562	2 408	52	477	2 833 (85)	204	
1974	2 461	2 245	57	373	2 561 (88)	210	
1975	2 381	2 227	106	177	2 298 (97)	207	
1976	2 055	1 886	88	188	1 986 (95)	202	
1977	2 049	1 980	41	181	2 120 (93)	208	
1978	2 167	2 096	34	509	2 571 (82)	228	
1979	2 207	2 081	27	549	2 603 (80)	268	30
1980	2 707	2 610	10	888	3 488 (75)	317	30
1981	2 968	2 872	1	801	3 672 (78)	312	30
1982	3 598	3 416	7	474	3 883 (88)	324	30
1983	4 637	4 586	68	223	4 741 (97)	342	30
1984	6 258	5 212	202	34	5 044 (103)	342	18
1985	4 147	4 318	349		3 969 (109)	322	16
1986	3 561	3 794	558		3 236 (117)		
1987	4 245		754	6		356	
1988	4 150		468	35			
1989	3 790		272	519			

Notes: a Apparent mill consumption calculated as purchases plus imports less exports. Numbers in parentheses represent the ratio of purchases to apparent mill consumption (%).

b Estimates presented in Sicular (1987: 34,77). Estimates for 1984 and 1985 are simple averages of the prices for north and south China.

Sources: *China Statistical Yearbook*, 1987, 1989; *Almanac of China's Foreign Economic Relations and Trade;* Sicular (1987).

The limited quantitative evidence that is available (such as Simmons, Trendle and Brewer, 1988) does appear to be consistent with price having a significant effect on supply. Casual empirical evidence on supply behaviour, at least since 1978, points to substantial price responsiveness. The increase in cotton production following the rise in cotton prices in 1980 is particularly marked and suggests that the response of individual commodities to changes in their prices can be very substantial. As is evident from Table 3.1, an increase in nominal cotton prices of around 50 per cent above their 1977 levels, together with the introduction of a 30 per cent above-quota price bonus for cotton, was associated with production more than doubling by 1984. The subsequent stagnation of production was associated with steady or declining cotton prices as the over-quota premium was wound back.

Also evident from the table is the fact that the rise in cotton prices followed the major increase in cotton imports in the late 1970s. The ability of the agricultural sector as a whole to expand by drawing additional resources from the remainder of the economy would undoubtedly be much less than the response of individual commodities. However, the apparent price responsiveness of individual commodities is of importance in assessing the likely implications of changes in the foreign exchange regime, since such changes could have marked effects on the relative prices of competing commodities which are importables, exportables or non-traded goods.

The increases after 1979 in agricultural product prices, and particularly in marginal (above-quota) prices, were not the only factors affecting output levels. Over the same period, the organisation of agriculture changed from a commune to a household responsibility system (Niu and Calkins, 1986). Under the household responsibility system, household income is much more strongly linked to the value of output than under the commune system, where remuneration was based on inputs, as measured by work points. Recent estimates suggest that this institutional change was a major source of productivity growth, accounting for perhaps three-quarters of the productivity gains in Chinese agriculture between 1978 and 1984 (McMillan, Whalley and Zhu, 1989). Lin Yifu (1990) sees the change in agricultural incentives as contributing 50 per cent of the agricultural output gains of the early 1980s.

The introduction of the contract system for agriculture in 1985 (Sicular, 1987) appears to mark a further step towards greater market orientation in Chinese agriculture. The replacement of compulsory procurement quotas with voluntary contracts, and the reduction in the number of prices paid for the same product, reduced problems such as quota under-fulfillment associated with the dual-track system. The introduction of this system was associated with a reduction in the price paid by the state for an additional (marginal) unit of output. This reduction in marginal output prices, together with increased prices for some inputs (Sicular, 1986) and a decline in state investment in agriculture (Watson, 1989) are likely to have influenced the reductions in output of some agricultural products after 1985.

The industrial sector

The post-1978 reforms have substantially reduced the role of central planning decisions in the allocation of resources in Chinese industry. In the past, there was widespread agreement that this sector remained less subject to price influences and more subject to planning decisions than the agricultural sector (Lin, 1988; Naughton, 1985; Chow, 1987: 57). Reforms in 1984 and afterwards appear to have largely increased the role of markets in this sector (Chinese Economic

System Reform Research Institute, 1987), so this conclusion may require re-evaluation.

While there appears to have been considerable decentralisation of decision-making in industry, this is largely to provincial or municipal governments, rather than to enterprises (Lin, 1988: 16). If these governments were simply to set production quotas in much the same way that the central government had previously done, then such decentralisation would not contribute to increasing price responsiveness. There appears to be a good deal of evidence, however, that these levels of government do respond to prices, either directly by passing price signals through to enterprises, or indirectly by taking into account price signals in their own planning and investment decisions.

Results from a survey conducted by the Chinese Economic System Reform Research Institute (1987) provide an indication of the extent to which industrial output had become subject to market price influences by 1984. At that stage, only 24 per cent of the gross output of firms was directly under mandatory plans and 57 per cent of sales were made under mandatory plans. However, almost 73 per cent of major raw materials were obtained through the planning system.

Zhang Shaojie and Zhang Amei (1987: 55) also point to very high shares subject to either mandatory or guidance planning in the production and marketing of raw and semi-finished materials. Even for these goods, however, approximately 20 per cent of total sales were made on the market by the enterprise itself, a ratio substantially higher than in many agricultural industries. Zhang and Zhang's results, reprinted in Table 3.2, also reveal that very low proportions of the activities of urban or rural collectively-owned enterprises were determined by planning decisions. The same study also points to relatively high growth rates of enterprises with a greater market orientation, implying that the share of the economy responding to market signals is likely to increase even in the absence of further explicit policy changes. This apparent price responsiveness of local authorities and enterprises can lead to economic costs when price signals are distorted in favour of particular activities such as consumer goods (see Wong 1985: 270, for instance), and contributes to the case for reform of the pricing system, as well as the foreign exchange system.

Table 3.2 Share of inputs, production and marketing subject to state planning for enterprises under different types of ownership, 1984 (per cent)

	State owned	Urban collectives	Rural collectives
Supply	86.8	6.5	3.0
Production	29.5	12.2	4.0
Marketing	71.0	4..0	

Source: Zhang Shaojie and Zhang Amei (1987:53).

The two-tier pricing system greatly increased the autonomy of individual enterprises and hence their ability to respond to price changes (Diao Xinshen, 1987). It appears, however, that the impact of price changes on industrial supply was frequently diminished by setting delivery quotas as a fixed percentage of output, rather than as an historically fixed amount. Diao Xinshen (1987: 47) contrasts this proportional two-tier price system in industry with the potentially non-distorting fixed quota system he sees as more prevalent in the agricultural sector. The seriousness of the distortion resulting from a proportional quota system depends upon the share of output sold at state prices. When the share of output sold at official prices is low, as in the case of collectives covered in Table 3.2, the distortion of price signals will be very slight.

The incentive problems resulting from proportionality between output and contract output have been reduced through the widespread introduction of the contracted managerial responsibility system (*Far Eastern Economic Review*, 8 September 1988: 132). While this system provides for contracts of between one and five years, the general practice appears, from field interviews, to be to set output for a period of around three years. This system may help to reduce the apparent inefficiency of many state enterprises in China, as well as to increase the price responsiveness of the state sector This system typically allows enterprises to retain profits above a certain threshold level, allowing the enterprise to obtain the full marginal return from its output. In the short run, such a system should be non-distorting although the level of taxation will influence investment levels in the longer term.

Clearly, the market is not the sole means of allocating resources in Chinese industry. In particular, where some inputs such as electricity are close to being 'essential' and are supplied only through a planning system, planners do retain some control over resource allocation, as noted by Perkins (1988: 615). Furthermore, markets for factors such as land and labour are poorly developed, and transferring these inputs between activities frequently requires complex transactions such as mergers between enterprises (Byrd and Tidrick, 1987).

Similarly, the movement of labour between areas, and particularly from rural to urban areas, is affected by a range of restrictions. Where the incentives are sufficiently strong, however, significant reallocation of factors has occurred. For instance, the rapid expansion of township, village and private enterprise has expanded employment in industry by moving industry to the labour force, rather than the reverse.

Byrd (1987) and Wu Jinglian and Zhao Renwei (1987) discuss the operation of the two-tier pricing system in some detail. They each conclude that the system has an important role in improving resource allocation, despite major problems involved, particularly in adapting the system over time, and with incentives for the maintenance of capital equipment. Byrd (1987), in particular, emphasises the strong tendencies inherent in the system for out-of-plan production to increase as a share of output, both as total production grows and as enterprises find ways to shift production from within-plan to out-of-plan categories.

From the evidence surveyed, it seems clear that the current structure of the Chinese industrial sector allows a substantial degree of price responsiveness, despite the continuing importance of planning in the sector. Together with the evidence previously surveyed for the agricultural sector, this suggests that price changes resulting from changes in the foreign exchange and foreign trade system can be expected to affect production and consumption patterns. This assessment does not mean that markets within China are providing the 'right' incentives, since the foreign trade and foreign exchange systems may cause major economic distortions.

Foreign trade

Prior to 1978 the Chinese economy was relatively closed, with decisions on exports made in the planning process in accordance with an assessment of the level of imports required for particular purposes. All foreign trade was channelled through the centralised Foreign Trade Corporations (FTCs), with minimal opportunities for direct interaction between producing enterprises and importers of Chinese exports.

While the foreign trade plan remains at the centre of China's foreign trade regime, its nature and implications have changed dramatically since 1979. The plan has been identified as having three major functions (World Bank, 1988b: 104):

1 protection of local production where it competes with imports;
2 mandating certain exports to offset the overall anti-trade bias of aspects of the trade regime; and
3 attempting to balance imports and exports.

The overall foreign trade plan involves both command plan and guidance plan components. Command plans are specified in physical units and are mandatory for the provinces or enterprises to which they apply. Guidance plans, by contrast, are specified in value terms and involve a substantially greater degree of flexibility. The proportion of total trade covered by command plans has been declining, particularly on the import side, although the decentralisation implicit in this change has been reduced by the introduction of measures such as import licensing, particularly since 1986.

The following account of China's export and import system is based heavily on the comprehensive World Bank (1988b) study, which provides a 'snapshot' of the system operating in late 1986. Other sources and interviews by the author have been used to update the assessment to late 1989.

The export regime

Despite substantial reform and liberalisation since 1978, the command plan system still covered 120 commodities, including coal, oil, agricultural products, textiles and garments and handicrafts in late 1986, and command plan exports accounted for an estimated 70 per cent of total export value (World Bank, 1988b: 22).

To some degree, the export controls were motivated by a desire to avoid adverse terms of trade effects in particular markets (Ho Lok Sang, 1986). While important in particular markets, such as food exports to Hong Kong, such terms of trade effects can easily be over-emphasised. Lin Shujian and Yang Yongzheng (1987) point out that, despite rapid expansion of its exports over the last decade, China's share of the total market in its major export markets remains quite low, even for its major labour-intensive exports, such as textiles and clothing. For instance, China's share of the total clothing market (in 1985) was less than 1 per cent in the United States, Japan and the European Community and 1.13 per cent in Australia.

Export controls on quota category exports of textiles and clothing to the United States, the European Community and other major markets are made necessary by the Multifibre Arrangement. However, the developing countries in general, and China in particular, have been able to expand their exports of textiles and clothing beyond the quota limits by exporting in unrestricted categories (Trela and Whalley, 1988: 20) and by exporting intermediate goods to Hong Kong. Given this, and the relatively low market share of China in these goods, there would not appear to be much justification for global export controls on the grounds that China is such a large exporter that it can increase its export returns by lowering the export volume.

In 1984 a major report on reform of the trade system by the Ministry of Foreign Economic Relations and Trade recommended a number of major reforms, some of which have begun to be implemented since 1986. The major recommendations included (World Bank, 1988b: 21):

1 direct participation by large producing enterprises in foreign trade;
2 eliminating the state's responsibility for profits and losses in foreign trade; and
3 adoption of an agency system for the FTCs.

The introduction of direct participation by selected producing enterprises in foreign trade was strongly endorsed by the World Bank (1988b) on the ground that the 'airlock' imposed by enterprises having to deal through the FTCs prevented producing enterprises from obtaining a great deal of information from buyers about production techniques and product quality (see also Chapter 7). Direct involvement of larger producing enterprises and consortia of smaller enterprises in foreign trade on their own account, however, appears to be increasing rapidly and has considerable potential to increase the efficiency and price responsiveness of the economic system.

A move towards achieving the second objective, made through the introduction of an experimental system for exports from several major sectors, was introduced in early 1988. Under this system, subsidies for exports were eliminated in the sectors of apparel, light industry, arts and crafts, and mechanical and electrical (Chan, 1989: 15). Instead, the FTCs are now able to retain a much higher percentage of the foreign exchange earned from exports in these categories (reportedly 75 per cent for apparel, arts and crafts, and light industry, and 100 per cent for mechanical and electrical).

One potentially major problem with these arrangements is related to their interaction with the domestic pricing arrangements. Since the prices of many goods in these categories have not been controlled, and since prices paid for items procured for exports are reportedly based on domestic prices, albeit with some negotiation, the pattern of exports could be distorted from that which would be observed if world prices were offered. Until recently, it was intended that this tension would be overcome fairly rapidly by phasing out domestic price controls. Recent concern about inflation appears to have put back the schedule for price reforms (*Australian Financial Review*, 3 October 1988: 26) and, as argued in the introduction to this volume, recent political developments in China may further delay or even reverse the process (*Far Eastern Economic Review*, 7 December 1989: 23). If the remaining price controls apply only to output up to a fixed quota level, they may have no effect on marginal responses and affect economic outcomes only through distributional effects, as discussed in the second section of this chapter. It remains to be seen, however, what form price controls will take over the next few years.

As long as two-tier or relatively free pricing is retained domestically, the new experimental provisions for foreign trade can be expected to have major implications for the performance of the export sector. The number of FTCs has increased substantially in recent years, and the restrictions on the scope of their activity have been relaxed. Thus price and non-price competition between these enterprises can be expected to make returns to enterprises tend towards the export market return adjusted for any incentives (such as foreign exchange retention) provided by government. While the strong traditional ties between FTCs and producing enterprises may limit the extent of price competition, the system for these major export sectors seems likely to be relatively responsive to the price incentives provided by the system.

The emergence of an agency system for exports, where the FTCs sell exports on behalf of enterprises in return for a specified commission, has been relatively slow, apparently because of the domestic pricing distortions for many goods, which make exports of most consumer goods unprofitable at the official exchange rate. By contrast, the artificially low prices of most producer goods make exports of these goods highly profitable and have resulted in the introduction of direct export controls. While the agency system is the intended long-term path of development for the export system, its full emergence seems likely to be retarded by the slowdown since late 1988.

Under the foreign trade system operating in 1986, it was estimated that 70 to 80 per cent of total exports were within the plan (World Bank, 1988b: 11). Since then, it seems likely that the share of above-plan exports has increased markedly. As noted in the World Bank study, the system prevailing at that time provided substantial incentives to expand out-of-plan exports. Further, the new experimental arrangements for clothing, arts and crafts, light industrial products and machinery and electronic exports greatly increase the extent to which the level of exports is price responsive. Given the importance of these export categories (Lin Shujian and Yang Yongzheng, 1987; World Bank, 1988b: 122), it now seems likely that 50 per cent or more of total exports are price responsive and hence directly affected by the exchange rate and other pricing policies, suggesting that price behaviour is now critical in determining trade outcomes. Indirect effects on the behaviour of local governments (World Bank, 1988b: 107) further increase the price responsiveness of the export sector. While some major export categories such as petroleum remain centrally determined, it is clear that the behaviour of the export sector can be affected substantially by price policy changes.

The import regime

The import regime appears to have undergone relatively little change in recent years. The key features of the system are:

- a command plan for imports of important raw materials — steel, chemical fertiliser, rubber, timber, tobacco, grain, polyester and other synthetic fibres (World Bank, 1988b: 111);
- central allocation of foreign exchange for imports on priority investment projects;
- allocations of foreign exchange for other priority imports of raw materials, spare parts and equipment; and
- non-centrally funded imports or imports subject to import licensing.

The World Bank estimated command plan imports to be around 40 per cent and non-centrally funded imports to be 30 to 40 per cent of total imports in 1986. From interviews by the author in June 1988, it appeared that central purchases of many commodities (and particularly wool) had declined markedly since that time, with a corresponding increase in local orders.

Non-plan imports are divided into restricted and non-restricted goods. Restricted goods, of which there were 45 in 1986 (World Bank, 1988b: 136), included most of the command plan imports (for instance, steel, rubber, timber, synthetic fibres and tobacco), 'luxury' consumer durables such as motor vehicles, televisions and refrigerators, and assembly lines for such 'luxury' consumer durables. In late 1986, in the wake of a serious foreign exchange shortage, the list of restricted imports was expanded to include wool.

The system of allocating some imports of raw materials at plan prices frequently involves implicit or explicit subsidies. However, since the quantities allocated are infra-marginal, subsidies to a particular industry will not affect short-run output decisions.

Import licensing imposes a total ban on imports of some goods perceived as wasteful—for example, motor vehicles—and sets very strict limits on goods already perceived to be in 'excess' supply, such as assembly lines for many consumer durables. Since the primary objective of import licensing is to keep the current account in reasonable balance, the extent to which it constrains imports will depend upon the seriousness of imbalances in the exchange rate and in domestic spending. For restricted goods under the import plan, an import licence can be obtained relatively readily (World Bank, 1988b: 138), while an application outside the plan is subjected to careful scrutiny to ensure, among other things, that the price paid by importers is not excessive. While the total imports by a province or enterprise, as well as imports of particular goods, are apparently constrained by import licensing, it appeared, in 1986 at least, that a restricted import with suitable finance, and for which no domestic substitutes were readily available, would eventually receive approval (World Bank, 1988b: 113). Under these conditions, imports are essentially being constrained by the availability or cost of foreign exchange rather than the commodity-specific import licensing system.

The import tariff includes a minimum tariff schedule and an import surcharge, whose combined rate ranges from zero to 200 per cent (World Bank, 1988b: 146). Where imports are currently banned, as has been the case for motor vehicles, the tariff is essentially irrelevant. Where total imports of a good are restricted by import licensing, the sole effect of the tariff is on the allocation of economic rents between the government and the licence holder. In all other cases, the tariff rate will affect the level and mix of imports.

The tariff structure includes several categories of exemption (World Bank, 1988b: 150), including:

- imports used directly for export;
- capital goods used for the transformation of industry; and
- imports of intermediate and capital goods for the Special Economic Zones, the coastal cities and Sino–foreign joint ventures.

The exemption for re-exports does not extend beyond the importing firm itself, and this appears to pose some problems for producers of intermediate goods (such as wool top-makers) who do not directly engage in exporting.

The overall pattern of the tariff schedule appears to be broadly consistent with the structure of import licensing, with those goods essentially banned under the import licensing system (such as motor vehicles) facing very high tariffs (200 per cent in the case of motor vehicles) and synthetic fabrics attracting quite high tariff rates (140 per cent). From interviews by the author, it appeared that most wool imports for domestic use attracted a duty of 20 per cent in 1988, while wool imported for processing and re-export was duty free.

For non-command imports, the FTCs usually act as agents for the purchasing enterprises. The enterprises are generally free to select an appropriate FTC (World Bank, 1988b: 23). Thus the price signals resulting from exchange rate changes and import tariffs would generally be transmitted back to the importing enterprise. An important exception to this principle arises where the resulting selling price is below the fixed domestic price, in which case the domestic price is charged. Thus imports of goods such as televisions are reportedly very profitable to FTCs, with the mark-up acting like a tariff except that the revenue accrues to the corporation rather than the government.

The import regime for wool has changed dramatically in recent years. Prior to 1988, the system was highly centralised, with all imports handled by eight corporations licensed to handle wool imports (Australian Wool Corporation, 1988). Wool purchases were financed primarily from allocations of foreign exchange by central and provincial governments.

In 1988 major reforms were instituted with the objective of transferring decision-making away from central governments to provincial departments and enterprises. The number of foreign trade corporations licensed to handle wool was increased, and some larger processors gained the right to import directly. Processing enterprises were able to purchase wool imports from their own foreign exchange earnings (Cross, 1989). From interviews by the author, it appears that central purchases of wool for distribution through the allocation system at fixed prices declined to 50 per cent or less in 1988.

The 1988 liberalisation was followed by a recentralisation in early 1989. Under the new policy, all wool imports became subject to control by a central agency, the China Central Wool Purchasing Group. While some larger enterprises still have direct importing rights, these are now subject to central approval. Under this regime, wool imports are therefore again subject to at least potential direct quantitative control, a control which appears to have been utilised in 1989 and 1990. Given the continuing desire to move away from quantitative controls and towards decentralised decision-making, an optimistic interpretation would be that the current tight control over imports will again be liberalised when the trade balance problems which caused the return to the use of quantitative controls become less pressing.

Summary

The foreign trade system has become considerably more price responsive and flexible than in the past. While command plan exports remain a substantial component of total exports, exports in general are now substantially responsive to the price signals generated by world markets and the exchange rate. The possibility that total exports have been artificially stimulated by excessive use of planning targets is less plausible, given that export subsidies for the large export sectors of clothing, light industry and arts and crafts have been replaced by relatively high retention rate arrangements for foreign exchange.

World market prices, and the exchange rate, have a strong impact on the demand for imports, given the general use of the agency system for imports. The effect of these forces is attenuated in some cases by import licensing, by tariffs and by interactions with official minimum prices, and more generally by the shortage of foreign exchange at the official exchange rate. Import tariffs allow the effects of changes in prices to be transmitted into the domestic economy and hence present no difficulty for assessing the effects of changes in policies such as foreign exchange retention arrangements. In principle, import licensing or quota systems insulate domestic prices from the effects of exchange rate changes and hence make these categories of trade unresponsive to exchange rate policy changes. However, these policy instruments are used primarily to balance the supply of and demand for foreign exchange, and their impact on behaviour can be analysed using standard economic methods.

Overall, it seems that the foreign trade system has become more flexible, more price responsive and probably more efficient than the more centralised system which it replaced. More importantly, the system is considerably more transparent than a pure planning system and hence inferences about the implications of exogenous shocks or policy changes can be made.

Nature of the foreign exchange system

Given the degree of price responsiveness of the major sectors of the Chinese economy, the foreign exchange system has an important role in influencing trade and the overall performance of the economy. The purpose of this chapter is to examine the key features of the Chinese foreign exchange system and to assess its likely implications for the traded goods sector.

Prior to 1979 the exchange rate had a relatively limited role in influencing the allocation of resources. With a heavy emphasis on planning in determining the level and pattern of exports, the exchange rate performed largely an accounting function, without influencing the pattern of exports and imports.

Since 1979 the foreign exchange system has been substantially reformed. A major change was the substantial devaluation, at least for trade-related transactions, involved in the introduction of the Internal Settlement Rate for Trade in 1981. The major 1981 reforms, however, also involved a highly centralised system of foreign exchange control with tight restrictions on holding of foreign exchange (Zhang, 1987). In 1985 the official exchange rate was essentially merged with the internal settlement rate and further devaluations occurred in 1986 and 1987, as is evident from Table 3.3. However, the exchange rate was then pegged to the US dollar until the 21 per cent devaluation in December 1989 (*Far Eastern Economic Review*, 21 December 1989: 39).

A particularly important recent change has been the introduction of a large number of legal secondary markets for foreign exchange (*Australian Financial Review*, 19 August 1988) in which enterprises with access to foreign exchange and the government itself can sell foreign exchange at a market determined price. These foreign exchange adjustment centres handle a large proportion of the foreign exchange available for transfer between enterprises, with an estimated one-third of retained foreign exchange being handled by these markets in 1988 (*Far Eastern Economic Review*, 2 March 1989: 103). The existence of these markets reduces the inefficiency arising where foreign exchange is rationed by non-market means and not all foreign exchange is used for high-priority uses (Dervis, de Melo and Robinson, 1981). Some degree of arbitrage between centres is possible and reportedly keeps the extent of divergences between provincial markets relatively small.

Zhang (1987) notes that the marked devaluation occurring in 1981 appeared to have a number of beneficial effects. In particular, he points to substantial growth in the level of exports, and increases in both the diversity of exports and the share of exports derived from outlying provinces. In view of the apparent effectiveness of earlier exchange rate adjustments in achieving these ends, and in reducing the need for export subsidies, the apparent slowdown in the adjustment of the exchange rate seems somewhat surprising. From interviews with officials in 1988 and 1989 it appears that concerns about the inflation associated with previous devaluations have inhibited further adjustment of the exchange rate.

There are a number of clear indications that the official exchange rate remains substantially overvalued relative to its equilibrium in the absence of foreign exchange controls. One strong indication is the 'shortage' of foreign exchange,

Table 3.3 China's official exchange rate and internal settlement rate for trade, 1975-89 (yuan/US$)

	Official exchange rate[a]	Internal settlement rate	Cost of earning foreign exchange
1975	1.86		
1976	1.94		
1977	1.86		
1978	1.68		
1979	1.55		
1980	1.49		
1981	1.70	2.80	2.61
1982	1.89	2.80	2.86
1983	1.98	2.80	3.22
1984	2.32	2.80	4.00
1985	2.94	2.80	5.00
1986	3.45		na
1987	3.72		na
1988	3.72		na
1989	3.72/4.72[b]		na

Notes: na — not available.
a Period average, yuan/US$.
b From December 21, 1989.

Sources: IMF, *International Financial Statistics*; Ho Lok Sang (1986); Tam Onkit (1987).

which necessitates a policy of strong controls on the use and holding of foreign exchange (Zhang, 1987). Such a shortage arises because, at the official rate of exchange, imported goods are artificially cheap, and selling goods on the export market is not sufficiently attractive to generate the volume of imports demanded.

Another indication of the overvaluation of foreign exchange has been provided by the substantially higher rates apparently prevailing in the secondary markets for foreign exchange. In August 1988 the rate on the Shanghai secondary market for foreign exchange was reported to be around 6.3 yuan/US dollar, well above the official rate of 3.72 yuan/US dollar (*Australian Financial Review*, 19 August 1988), and Shan (1989: 48) reports a secondary market peak of 7.06 on 9 November 1990. By late 1989 the secondary market rate had declined substantially, to between 5 and 6 yuan/US dollar in response to reductions in spending and direct curbs on imports (*Far Eastern Economic Review*, 21 December 1989: 39). Since transactions in these secondary markets are a legal means of transferring foreign exchange between entities, these rates do not reflect the risk premium likely to be associated with black market transactions. Rather, such high rates in the secondary market would appear to reflect a very marked overall shortage of foreign exchange.

On the basis of the evidence on price responsiveness presented earlier, it can be assumed that the supply of foreign exchange is an increasing function of the price of foreign exchange. Similarly, the demand for foreign exchange can be expected to be a downward sloping function of its price, with a higher price discouraging use of foreign exchange for imports and diverting purchases to domestic sources of supply. Given these assumptions, the market for foreign exchange can be represented by the 'short-side disequilibrium model' presented in Figure 3.2.

The figure corresponds to the suppressed balance of payments deficit case considered by Desai and Bhagwati (1979) and is a model which appears appropriate in the light of China's success in avoiding excessive trade deficit problems throughout most of the 1980s. In this general class of model, the overvaluation of the official exchange rate is maintained by some form of restriction on imports constraining the total value of imports to equal the value of exports (plus borrowing) in this model.

In the simple short-side rationing model depicted in Figure 3.2, the supply of foreign exchange is determined by the official exchange rate and the domestic supply curve SS. At this rate, the demand for foreign exchange exceeds the supply. The demand for the available quantity, q_o, then determines the secondary market price of foreign exchange $\acute{e}$. At the aggregate level, the volume of imports can be viewed as being constrained by the supply of foreign exchange. From the perspective of individual enterprises able to buy or sell foreign exchange, the constraint on imports is the high price of foreign exchange $\acute{e}$. The result of this two-tier pricing system for foreign exchange is a reduction in imports from q^* to q_o.

The simple diagram presented in Figure 3.2 indicates only the effects of the foreign exchange management system. Direct foreign trade interventions such as tariffs, export taxes, import quotas and import licensing influence the position of both the supply and demand curves for foreign exchange. Overvaluation evident in the official exchange rate and the undervaluation of the secondary market rate are relative to the equilibrium exchange rate in the presence of these distortions, e^*. Because of the trade distortions, as well as foreign borrowing decisions, e^* may be above or below the long-run free trade equilibrium rate.

At first sight, the model represented in Figure 3.2 might appear to be a partial equilibrium representation of a general equilibrium phenomenon. However, it can relatively easily be derived from a simple general equilibrium model such as that developed by de Melo and Robinson (1989). A more detailed discussion of the underlying basis for Figure 3.2 is given in Martin (1990c).

The simplified representation of exchange rate determination presented in Figure 3.2 omits some key features of the market for foreign exchange. The first is the extent to which enterprises, or the provincial authorities responsible for enterprises, are entitled to retain a certain proportion of their foreign exchange earnings. The effect of this foreign exchange retention scheme is to shift the supply curve of foreign exchange to the right over that portion of the curve for which foreign exchange remains in shortage. As long as the official rate is below e^*, foreign exchange will command a premium, and the potential value of retained foreign exchange earnings, either directly for purchasing inputs or through their sale on the secondary market, provides an incentive for enterprises

Figure 3.2 Supply, demand and price of foreign exchange

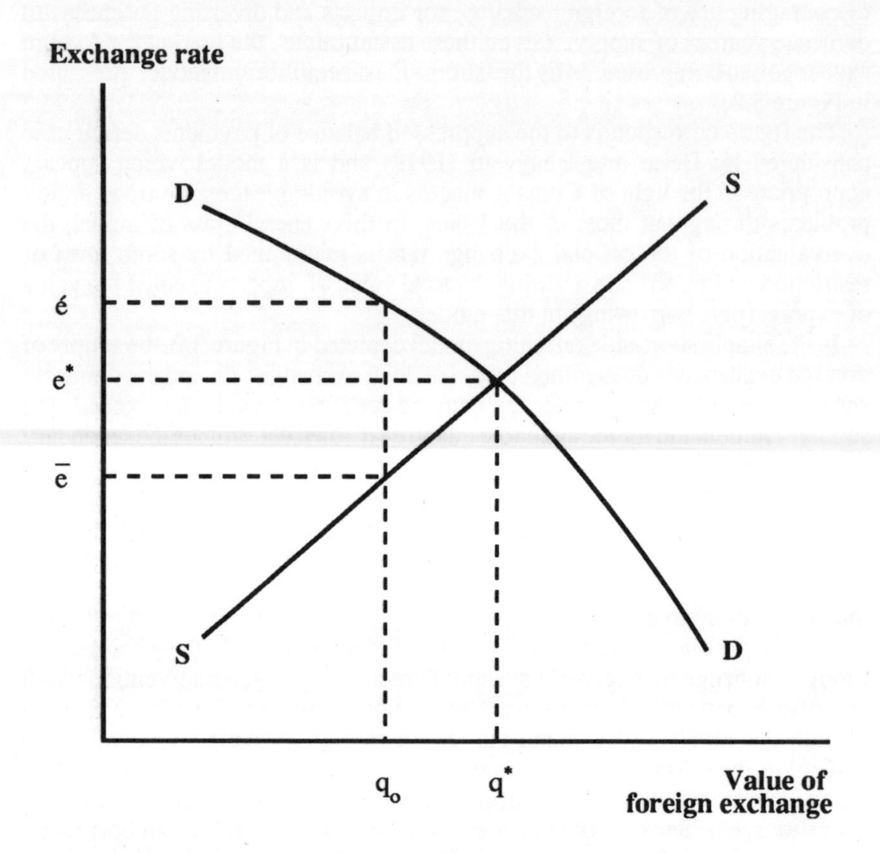

to increase their exports. Thus in Figure 3.3 the new supply curve of foreign exchange S'S is drawn to the right of SS at all exchange rates below e*.

As is evident from Figure 3.3, the effect of a foreign exchange retention arrangement is to increase the supply of foreign exchange and to drive the secondary market price down from e^1 to e^2. Tam Onkit (1987: 9) estimates that the average retention rate of local authorities, ministries and enterprises was around 30 per cent in 1986, while Zhang (1987: 52) points to a wide range of retention ratios for different products and regions, with most commodities appearing to be in the range 25 to 100 per cent.

Where the base rate of 25 per cent retention applies, the usual practice is for 12.5 per cent to be allocated to the enterprise and 12.5 per cent to the provincial government. Even with a relatively large divergence between the official exchange rate and the secondary market rate, such a retention rate provides only a limited incentive to exporters. It appears, however, that provincial governments frequently respond to this incentive by encouraging the enterprises under their control to expand their production for export. Where provincial govern-

Figure 3.3 Supply, demand and price in the presence of a foreign exchange retention scheme

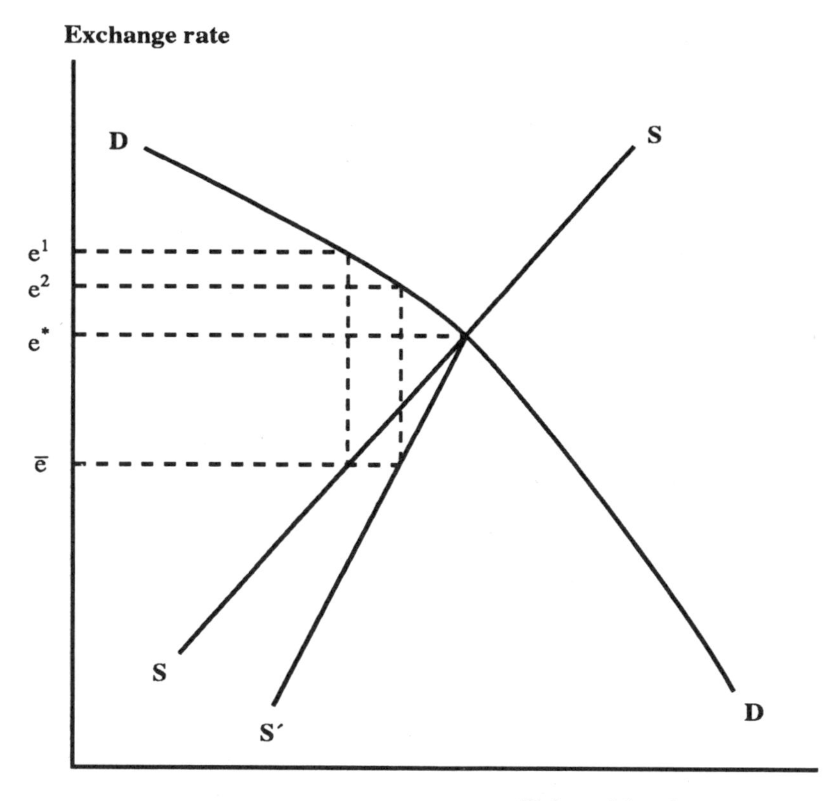

ments react in this way, the effect of the foreign exchange detention arrangements may be greater than would be indicated by the direct retention rate to the enterprise.

As was noted in the second section of this chapter, the retention rate applying to a substantial proportion of trade is probably now substantially above the base rate of 25 per cent. The introduction of the experimental system under which exporters in the arts and crafts, clothing and light industry sectors receive a retention rate of 70 or 80 per cent means that some of China's most important exports are now covered by this much higher retention rate (Chan, 1989; Shan, 1989). While this system generally applies to the FTCs, rather than directly to the producing enterprises, the increased ability of enterprises to choose between FTCs, and the possibility of negotiation between the FTC and the enterprise over a range of factors such as price and the supply of material inputs, presumably means that price signals are transmitted fairly effectively.

A foreign exchange contract system also allows some enterprises to obtain a higher rate of foreign exchange retention for above-plan exports. The introduction of this system, and the experimental system of higher retention rates, has been associated with a reduction in the extent of distortions in both product and input markets. Under both systems, it appears that export subsidies are not available on marginal units of production. Furthermore, the higher rates of foreign exchange retention have been associated with reductions in the central allocations of material inputs.

By increasing the incentive to export, and hence increasing the available supply of foreign exchange, the retention arrangements can be expected to bring the secondary market price closer to the official price, as is evident in Figure 3.3. If the retention rate were raised to 100 per cent, the secondary market rate would be equal to the equilibrium rate. The equilibrium rose (a devaluation) substantially in 1988 because of the high rate of inflation. However, it appears to have fallen substantially in late 1989, following the introduction of controls on credit and spending, and tightening of the import licensing system.

IMPLICATIONS OF THE FOREIGN EXCHANGE SYSTEM FOR OVERALL TRADE

The major implications of China's foreign exchange system are those arising from the overvaluation of the currency and the consequent disincentive to export. In addition, it creates an incentive for individuals to devote time and resources to non-productive activities such as attempting to increase their access to artificially scarce foreign exchange. Because the nominal value of the official exchange rate is fixed, changes in the money supply and hence in the general price level change the relative prices of exports and other goods, rather than merely the average level of prices. As a consequence, major economic distortions can arise without explicit policy action.

The most direct effect of an overvalued exchange rate is to lower the price obtained for exports. If the price of an export good on the world market is P^*, the price received by a domestic enterprise for a unit of the good exported is eP^*. If the exchange rate is overvalued, e is too low and hence eP^* is reduced accordingly. Devaluation of the currency increases e and hence increases the domestic price of the good.

The increased price of exports following devaluation of the official exchange rate raises the supply of exports. If China is a price-taker on world markets for a particular commodity, this increase in the supply of exports will increase the foreign currency value of exports. In fact, as long as the world demand for a particular export has a demand elasticity greater than one in absolute value, the increase in exports will increase the supply of foreign exchange. Devaluation will also tend to increase the flow of foreign exchange from other sources, such as remittances.

A second consequence of overvaluation follows from its negative effect on the supply of foreign exchange. Since the total available supply of foreign exchange is reduced, total expenditure of foreign exchange must be correspondingly reduced. This effect is brought out in Figure 3.2, where the total supply of, and demand for foreign exchange, is reduced from q^* to q_o. While one of the justifications frequently offered for maintaining an overvalued exchange rate is to encourage the import of capital and productive inputs to facilitate development, the effect of overvaluation must be to reduce total imports. Any purchases of capital or intermediate goods must therefore be made from within a diminished total supply of foreign exchange, and hence are likely to face increased competition from other claims on foreign exchange. Unless a particular category of imports can obtain a substantially larger share of the total available foreign exchange than it would in an undistorted situation, the level of its imports must decline.

As was noted in the previous section, if foreign exchange is allocated between the competing demands on the basis of the returns available in each use, a scarcity premium will develop. The cost of foreign exchange to an enterprise will be $\acute{e}$, above the equilibrium exchange rate e^*, which in turn is above the official exchange rate $\bar{e}$. Where an enterprise has access to a secondary market for foreign exchange or 'foreign exchange adjustment centre', this scarcity premium becomes explicit, either as an increase in the cost of purchasing imports or as the opportunity cost of using foreign exchange when the scarcity premium ($\acute{e} - \bar{e}$) could be obtained by selling it in the market. If enterprises do not have access to a secondary market, the scarcity premium is just as real, although it will vary substantially between enterprises.

As Dervis, de Melo and Robinson (1981) have demonstrated, the effects of the foreign exchange overvaluation on the economy will depend on whether scarce foreign exchange is allocated using a secondary market or some other allocation rule. If a secondary market is used, the value of foreign exchange, at the margin, will be the same in all cases and hence the available foreign exchange will be used efficiently. By contrast, if a relatively arbitrary but plausible rule is used for allocating the scarce foreign exchange, such as equi-proportional reductions in all import demands, the overall efficiency losses will be greater.

The implicit tax on exports implied by an overvalued exchange rate reduces the size of the export sector. There is also an implicit tax on imports, however, which has the opposite effect on the import-competing sector. The scarcity of foreign exchange created by overvaluation leads to higher prices of foreign exchange, which means that production of import substitutes will expand and consumption of importables will decline. Many goods which could be imported at lower cost are, instead, produced domestically.

The political pressures to stimulate imports of investment goods and raw materials, which contributed to the original decision to overvalue the currency, are likely to further stimulate the domestic production of 'non-essential' or 'luxury' goods by reducing the share of imports allocated to such goods. It seems likely that this has happened in China. Wong (1985: 270) points to very high profit levels in watchmaking, for instance. This factor may also contribute to the widespread concern amongst Chinese authors about the apparent bias towards consumer goods, at the expense of heavy industry, in recent years (Chinese Economic System Reform Research Institute, 1987: 12). Once production of substitutes for importable consumer goods is established, its demands for raw materials and parts add to the overall demand for foreign exchange, intensifying the overall shortage of foreign exchange. Measures such as the use of import licensing to prevent the import of additional assembly lines may help to reduce the severity of this problem, but if the profit incentive remains, they are likely to result in even higher cost responses, such as domestic production of assembly lines.

Although the overvalued exchange rate regime stifles the production of exportables, and stimulates the production of importables, it has exactly the opposite effect on consumption of these commodities. Domestic consumption of goods which continue to be exported is stimulated by their low price, while domestic consumption of importables is discouraged by their artificially high price. Given the apparently large degree of overvaluation of the yuan, these relative price effects are potentially important in markets such as the fibre market, where importables such as wool and synthetics compete directly with an exportable such as cotton.

IMPLICATIONS FOR THE FIBRE AND TEXTILE MARKETS

It appears that the major administrative and policy influences are very different between the fibre and textile/clothing sectors and so the characteristics of each have been considered in turn.

The market for fibres

Up to 1984 the market for wool was relatively tightly controlled, with the price paid to producers determined by the state and sheep numbers and wool output levels subject to central controls. After 1982 households gained ownership of their flocks and hence wool supply can be expected to have become much more responsive to the prices of both wool and sheep meat. Abolition of the unified purchasing system at the beginning of 1985 sparked a 'wool war' which resulted in substantial increases in wool prices (see Chapter 9 in this volume).

As was noted in an earlier section, the supply of cotton in China appears to be responsive to prices paid by producers. While the price of this commodity is subject to official control, there is evidence of substantial 'illegal' premiums being paid by purchasers, suggesting that there is some responsiveness of the actual prices paid to market conditions (*Far Eastern Economic Review*, 2 March 1989: 83).

The market for synthetic fibres appears to be heavily controlled, with imports of synthetic fibres and the chemicals used to produce them strongly restricted. The objective of this policy appears to have been to protect the synthetic fibre industry, in which China is unlikely to have a comparative advantage. The tariff structure provides for very high effective rates of protection to these activities (World Bank, 1988b). Recently, however, it has been reported that retail prices for synthetic fabric have been reduced, perhaps reflecting some reduction in emphasis on this objective, or the development of a more efficient domestic chemical fibre sector.

Table 3.4 Raw wool production, consumption and trade, 1978-89 (kt, clean weight)

	Production	Raw wool imports	Net domestic consumption[a]
1978	65.91	11.529	77.4
1979	66.373	12.683	79.1
1980	72.082	20.901	93.0
1981	70.562	37.744	108.3
1982	81.410	55.790	137.2
1983	74.710	74.060	148.8
1984	77.000	75.000	152.0
1985	72.000	91.000	163.0
1986	66.000	110.000	176.0
1987	70.000	113.000	185.0
1988	74.253	138.700	213.0
1989	79.713	77.25	156.9

Note: a No allowance is made for changes in stocks, and wool top imports are not included.

Sources: Table 6.7, and *China Statistical Yearbook*, 1989.

Demand for wool increased continually from 1978 to 1988, as is evident from the data presented in Table 3.4. The main factors underlying this expansion appear to have been the rapid growth in incomes in China in recent years, increasing urbanisation, and changes in styles of clothing (Angel, Simmons and Coote, 1988; and Chapter 2 in this volume). While domestic production increased between 1978 and 1988 this growth was uneven and substantially below the growth in consumption of 175 per cent over the same period (see Table 3.4). Wool production actually peaked at 81.4 million kg in 1982 (clean weight) and has since declined to 66 million kg in 1986, rising to just below 80 million kg in 1989. Imports of wool rose from 11.5 million kg in 1978 to 139 million kg in 1988, before plummeting in 1989. Imports have been above 20 million kg since 1980 and exceeded domestic production from 1985 until 1988. Based on the figures presented here, wool production appears likely to be an import competing industry for some time. By contrast, the wool textile industry is export-oriented, with a substantial volume of wool contained in fabric and garment exports.

Basic data on the cotton market were presented in Table 3.1. The most pronounced change evident in Table 3.1 was the dramatic increase in raw cotton production between 1979 and 1984, when cotton production virtually tripled. Also of interest is the decline in production after the exceptional 1984 harvest, to levels around those prevailing in 1982 and 1983. With expansion of the textile industry, estimated mill consumption of cotton increased dramatically up to 1984 but then appears to have declined in 1985 and 1986. The extent of any decline in consumption during this period is almost certainly exaggerated. Stock levels were very high after the bumper harvest of 1984 and allowed domestic consumption to remain temporarily above estimated levels. Consumption levels have risen relative to production, with the result that China became a net importer of cotton in 1989.

Given the model of the foreign exchange market presented in the previous section, China's return to being an importer of cotton after 1989 would have been expected to place substantial upward pressure on the price of cotton. This pressure on prices has been partially resisted, with black market premiums emerging (*Far Eastern Economic Review*, 29 March 1989: 15), and users without access to black market cotton supplies have been constrained in the quantity of cotton they can use.

Some basic statistics on chemical fibre production, imports and total availability are presented in Table 3.5. From these figures it is clear that total chemical fibre production has been rising rapidly. Imports have also risen rapidly, but not as fast as production, and in 1989 accounted for about 25 per cent of consumption. Given the fact that synthetic fibre production is a highly capital-intensive process in which China is unlikely to have a comparative advantage, it seems likely that China will remain a net importer of synthetic fibres for some time.

Table 3.5 Production, imports and availability of chemical fibres, 1975-89 (kt)

	Production	Imports	Availability
1975	155	108	263
1976	146	169	315
1977	190	205	395
1978	285	268	553
1979	326	215	541
1980	450	411	861
1981	527	629	1 156
1982	517	434	951
1983	541	398	939
1984	735	530	1 265
1985	948	831	1 779
1986	1 017	475	1 492
1987	1 175	370	1 545
1988	1 301	535	1 836
1989	1 465	503	1 968

Sources: *China Statistical Yearbook*, 1986, 1989 and earlier issues; *Almanac of China's Foreign Economic Relations and Trade*, 1984:7; *China's Customs Statistics*, various issues.

Data on the purchase prices of wool and cotton are presented in Table 3.6 to provide some information about price incentives for the production of cotton and wool in China. As is evident from the table, the purchasing price of wool (on a clean basis) was around three times the price of cotton in the 1978 period. When the purchasing price of cotton was increased in 1980, however, the price of wool remained virtually unchanged and the ratio of the price of wool to the price of

Table 3.6 Purchase prices of raw wool and cotton in China, 1975-88

	Greasy wool (yuan/kg)	Cotton (yuan/kg)	Ratio of wool to cotton price (clean price)[a]
1975	3.1	2.1	3.0
1976	3.1	2.0	3.1
1977	3.3	2.1	3.2
1978	3.4	2.3	3.0
1979	3.4	2.7	2.5
1980	3.4	3.2	2.2
1981	3.5	3.1	2.2
1982	3.6	3.2	2.2
1983	3.7	3.4	2.2
1984	3.7	3.4	2.2
1985	5.0	3.2	3.1
1986	6.0	3.2	3.7
1987	6.3	3.4	3.7
1988	10.8	4.0	5.4

Note: a Assuming a clean yield of 50 per cent for wool.
Source: *China Statistical Yearbook,* 1987: 664, 1989:708.

cotton fell to around 2 yuan until wool prices rose in 1985, returning the wool/cotton price to around 3 yuan. The increase in wool prices in 1985 and 1986 following the price reforms initiated in 1985 was a consequence of the commodity marketing reforms of 1984 (see Chapter 9). By 1989 the producer price of wool had risen dramatically relative to the price of cotton. The changes in the producer price of wool relative to cotton are consistent with developments in production. Cotton output grew in the 1977–84 period when its price rose relative to the wool price and stagnated in the 1985–88 period when its price fell relative to the price of wool. The increase in the producer price of wool since the marketing reforms in 1984 appears to have stimulated production from the low level it reached in 1984.

Since wool was imported, and cotton was a net export item from 1984 until the poor harvest of 1989, the domestic wool to cotton ratio would generally have been expected to be above the international price ratio for commodities of the same quality because of the overvaluation of the currency. This ratio appears to have risen substantially since 1985 as the official purchasing price of wool has risen nearer to market prices.

The textile market

In contrast with the situation at the fibre level, it appears that the retail price of woollen fabric is, by international standards, very much higher than the retail price of cotton fabric. Even allowing for the greater width of the woollen fabric relative to cotton (1.5 m rather than 1 m), the difference between the two prices is striking. With the raw wool/raw cotton price ratio around 3 to 1, a 10 to 1 price ratio for wool and cotton cloth at the retail level requires explanation. While wool requires substantially more processing to convert it from the raw state to finished fabric, it seems unlikely that these additional costs could explain the greater divergence between the fabric prices. From interviews with wool and cotton processors in Shanghai and Guangzhou, it appears that labour costs make up only a small proportion of total costs (see also Chapter 6). A relatively high retail price for wool would be easily explained if wool products were importables, and cotton products exportables. However, as is made evident in Chapter 6, China remains a substantial exporter of wool products.

Another indication that domestic prices of wool fabric may have been kept high relative to export prices can be obtained by examining the export unit value series. For the limited time period over which data are available (1982-86), it appears that the per metre price of worsted wool fabric exports has been substantially below the average retail price of woollen and worsted fabric (see Table 3.7). Part of the difference between the two series could be due to retailing costs, although these are likely to be relatively low in a low labour cost country such as China. Another contributing factor is the rebate of the tariff on imported wool allowed for wool used in export products. However, the use of worsted fabric data for export unit values raises the unit value of exports relative to the retail price series, which includes generally lower priced woollen fabrics. Probably the most likely explanation is the existence of export incentives not reflected in the unit value data.

In contrast with wool, the export unit values for cotton are in most cases above the retail price. This is consistent with the average price of cotton fabric, a basic consumer staple, being held below the returns available on export markets. Another possible explanation would be that these export unit values are increased by quota rents on binding Multifibre Arrangement (MFA) quotas, and that marginal export returns obtainable from non-MFA markets are somewhat lower. This argument would also apply to exports of woollen fabric unless quotas on woollen fabric are non-binding.

There are several possible explanations for the volume of exports of wool in apparel remaining substantial while the domestic price is well above export returns. Perhaps the most likely cause is the relative loss of fibre identity at higher levels of processing, where value added becomes increasingly important as a share of the total value. This explanation would be consistent with a much higher share of wool in garment exports than in fabric exports. This hypothesis appears to be consistent with the actual pattern of exports. Although wool accounts for only 0.4 per cent of Chinese fabric exports, it makes up 4 per cent of garment exports (Australian Wool Corporation, 1988: 14). Given the importance of other inputs in the production process, and the differences between final products, an increase in the price of wool would be expected to reduce its share of the total export market, but not necessarily drive it out of the market altogether.

Other regulatory considerations probably also contribute to the continued presence of wool exports despite such an unfavourable export/domestic price ratio. One such factor is the partial separation between domestic and export production sectors. Wool imports for domestic use are subject to import

Table 3.7 Retail prices and export unit values for wool and cotton fabric, 1975-86 (yuan/m)

	Wool retail price[a]	Wool export unit value[b]	Cotton retail price	Cotton export unit value	Wool/cotton retail ratio
1975	15.5		1.6		9.7
1976	16.2		1.6		10.0
1977	16.0		1.6		10.0
1978	16.7		1.6		10.4
1979	17.1		1.6		10.7
1980	18.1		1.6		11.4
1981	18.6		1.6		11.8
1982	18.5	12.8	1.6	1.7	11.6
1983	18.1	11.9	1.8	1.6	10.0
1984	20.3	12.5	1.6	1.8	12.8
1985	22.8	13.0	1.7	1.8	13.4
1986	25.4	12.9	1.8	2.2	14.1

Notes: a Average for woollen and worsted fabrics.
b Worsted fabric.

Sources: *China Statistical Yearbook*, 1987; *Almanac of China's Foreign Economic Relations and Trade*, 1987.

licensing and customs duties, while imports of wool for production of exports are exempt from duty and unconstrained by import licensing. Presumably, enterprises which are able to retain all or part of their foreign exchange earnings and convert them to yuan on the secondary market find exports profitable. Another possible explanation for continuing exports over the period for which data are available was requirements that otherwise relatively autonomous enterprises maintain some sort of foreign exchange balance in their activities.

Although a number of possible reasons for the apparent high domestic price of wool fabrics have been advanced, the available information does not allow any inference about the previous structure of domestic prices relative to export prices. The direction of change in pricing and marketing reform has been an increase in the use of market prices to influence production and consumption decisions. Given the substantial change which has already occurred, it is reasonable to examine the behaviour of a stylised post-reform economic system in which decisions at the margin are guided by secondary market prices.

QUALITATIVE ASSESSMENT OF THE EFFECTS OF A CHANGE IN EXCHANGE RATE POLICY

The price analysis in this chapter highlighted, firstly, the increase in the level of responsiveness in the Chinese economy. This applies in all sectors, not only agriculture, and it was also argued that decisions to trade in the international market are becoming more sensitive to relative prices. Clearly, the price responsiveness of the Chinese economy is inhibited to a significant degree by some planning interventions. However, the degree of influence of prices seems sufficient to support the use of models of behaviour which focus on the effects of market forces.

A somewhat different two-tier pricing system operates in the market for foreign exchange with an official and a secondary market price. The official exchange rate is overvalued and, as a result, rates in secondary markets are higher than the official rate. In fact, it was shown that this overvaluation of the official role can be expected to lead to undervaluation of the secondary market rate. Conditions in the foreign exchange market, especially the use of foreign exchange retention rates of less than 100 per cent, suggest that deregulation of the market, by dismantling the system of foreign exchange controls which allow the overvaluation to be sustained, would actually lower rates in secondary markets. That is, a devaluation of the official exchange rate should, other things being equal, lead to an appreciation of the secondary market rate.

These results have a number of implications for the fibre and textile markets. Applying the simple model of the relevant markets allows us to undertake an experiment designed to make an initial assessment of the effects of devaluation. Deregulation of the foreign exchange market would be expected to lead to the following effects:

1 *A fall in the price of domestic raw wool and in domestic wool production.* Since wool production is an import competing industry in China, the scarcity premium on foreign exchange makes its price higher than in the situation of foreign exchange equilibrium. A devaluation, by reducing the scarcity of foreign exchange, would make wool cheaper to domestic users. This result assumes that the authorities react to the increase in the availability of foreign exchange by uniformly relaxing the import licensing regime which, after all, was introduced as a means of dealing with this problem.

2 *A rise in the price of exportable wool garments and textiles.* The rise in the price of exports following a devaluation would stimulate the diversion of garments from the domestic to the export market, raising the prices of these garments on the domestic market. While the effect of this development on China's demand for wool for use in export processing would be favourable, the negative impact of a price rise on domestic demand would reduce the global demand for wool. However, in practice, the domestic price of wool apparel appears to be very high and a reduction in the price of wool imports, associated with liberalisation, might place some downward pressure on the domestic price of wool garments. Furthermore, the domestic demand elasticity for garments is likely to be much lower than the demand elasticity facing China's exports on world markets. The fact that prices of all types of garments (and, indeed, all exportables) are expected to rise would ameliorate the effect of the rise in the price of wool garments. Otherwise, substitution between wool and other garments which were close substitutes would reduce demand substantially. It seems likely that the price elasticity of demand for apparel as a whole is relatively low.

3 *A change in the price of cotton.* When China is a substantial exporter of cotton a devaluation of the official exchange rate might be expected to raise the cost of cotton to users. This would be beneficial for the suppliers of substitutes such as wool and synthetics. By contrast, when China is an importer of cotton, as in the early 1980s, and more recently, the secondary market exchange rate appreciation resulting from devaluation of the official rate would lower the price of cotton, making it more competitive with other fibres. Thus the effects of devaluation on the competitiveness of cotton depends on whether China is an exporter or importer in the period under consideration.

4 *A fall in the price of synthetics.* Since this industry is import competing, an increased availability of foreign exchange and a lower scarcity premium following a devaluation would lower import prices and hence the price of domestically produced fibre. The fall in synthetic prices would be a negative factor for wool, although synthetic fibres are now used extensively in wool blends (see Chapter 6), possibly reducing the extent of competition between the two fibres.

5 *An increase in Chinese demand due to an increase in Chinese incomes and living standards.* Incomes would rise because of an expansion of trade as China increased the extent to which it gained from trade by increasing exports in which it had a comparative advantage and made greater use of the lower costs of other imported goods. This would be a strong positive factor, given the apparently very high income elasticity of demand for wool in China (Angel, Simmons and Coote, 1988).

In the five causal channels considered here, the first four depend only upon the changes in exchange rates brought about by liberalisation. The last depends upon relatively less direct changes in income levels and wage rates, brought about by liberalisation but not captured in simple diagrams such as Figure 3.2. The exercise of identifying causal channels reported in this chapter also highlights the potential importance of substitution effects between individual fibres and of those input demand linkages between fibre, textile and apparel industries not incorporated in the aggregative frameworks discussed in earlier chapters.

The overall impact on world wool demand of these policy changes in China would depend upon all of these impacts. The reduction in Chinese wool production would be a positive factor and the rise in the price of wool products in the Chinese market would be a negative factor, while the increased supply of Chinese exports onto world markets would lower the average costs of transforming wool into garments and hence tend to expand global wool consumption. The substitution effects between wool and cotton, and between wool and synthetics, would probably be very largely offsetting.

SUMMARY AND CONCLUSIONS

The qualitative analysis reported in this chapter covered the major features of the Chinese economy relevant to any evaluation of the effects of the foreign exchange system on the market for wool. Since the allocative effects of a pricing instrument such as the exchange rate depend upon the operation of the internal pricing system, the operation of China's two-tier pricing system was first surveyed. This led to the conclusion that China's pricing system does play a major role in resource allocation, with secondary market prices being particularly important.

Secondary market prices also play an important role in the foreign exchange market, although the price signals transmitted have been substantially distorted by overvaluation of the exchange rate. The simple model used to analyse the supply of demand for foreign exchange clarified the link between the official exchange rate and the secondary market rate, with overvaluation of the former leading to undervaluation of the latter in an overall context of foreign exchange 'shortage'.

The effects of exchange rate overvaluation on the market for wool were seen as depending on the whole array of relative price changes which it induces, including the relative price of wool and other fibres, the prices of textiles and clothing and the reduction in incomes brought about by overvaluation. At least the direction of some of these effects can be determined using simple back-of-the-envelope approaches. For instance, a devaluation can be expected to increase the availability of importables such as chemical fibres, and increase the domestic demand for apparel.

However, there would be some offsetting effects such as an increase in domestic production of wool, and so the overall direction of effect cannot be unambiguously determined using informal techniques such as those employed in this chapter.

A computable general equilibrium model of the Chinese economy is one means by which the range of interacting forces identified in this chapter might be analysed. The development of such a model and its application to this problem is the subject of the next chapter.

4 Effects of foreign exchange reform on raw wool demand: a quantitative analysis

WILL MARTIN

INTRODUCTION

In the previous chapter, an initial attempt to evaluate the effects of devaluation of the official exchange rate was made. While this attempt, based on an analysis of the major structural features of the market, generated a number of insights into the effects, it did not allow the direction or the magnitude of the effect on wool demand to be determined since it was not possible to determine the relative strength of the positive and negative forces.

The simple model used in the previous chapter also assumed that all products, even textiles and clothing produced by different countries, are homogeneous. However, there is considerable evidence that such product aggregates cannot satisfactorily be treated as homogeneous (Isard, 1977) and that a framework which incorporates differentiated products (for example, Armington, 1969) is required.

A numerical model which allows evaluation of the relative strength of competing forces, and which incorporates the differentiated nature of products such as textiles and clothing, is developed in this chapter. For those readers not interested in the modelling issues, the results of the analysis are summarised in the last section of the chapter.

This section is devoted to a schematic consideration of the major features of the system to be modelled. Following this, the second section contains a discussion of some of the major modelling issues involved. Broad features of the model are set out in the third section, data and structure in the fourth, equations of the model in the fifth, and simulation results are presented in the final section.

The system under study can be divided into three major components: China's macroeconomy, the Chinese textile and apparel sectors, and the world markets for textiles and apparel. These three broad components of the system are depicted in Figure 4.1.

The first component of the system (A) is China's macroeconomy. The price of imports relative to both exports and to non-traded goods is determined within this system. The system as depicted can determine only relative prices. If, however, some means of determining the absolute price level is introduced, then all of the nominal prices in the economy can also be determined.

Part B of Figure 4.1 deals with the textile/apparel market. Within this market, the three major classes of fibre inputs are combined with other inputs (materials, labour, capital) to produce textiles. Relative prices (including exchange rate effects) influence the allocation of these textiles between these two markets. Domestic textiles, and some imported textiles, are combined with other inputs to produce apparel. The relative prices obtained from the domestic and export markets then influence the allocation of apparel between the domestic and export markets.

Figure 4.1 Inter-related systems represented by the model

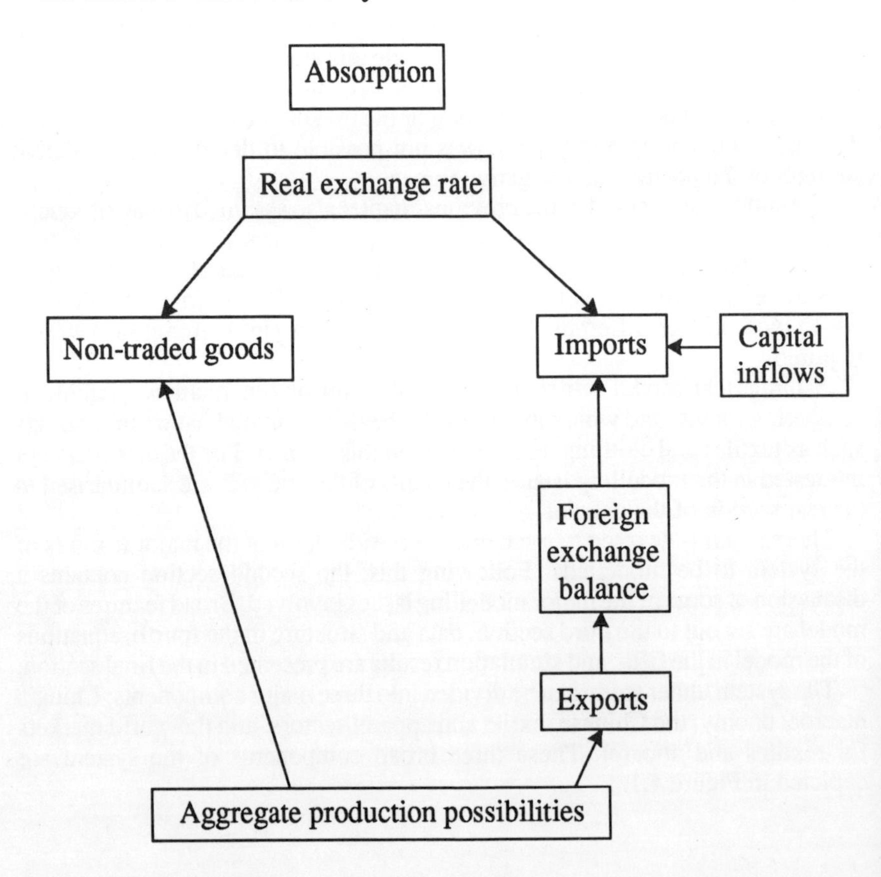

Figure 4.1 (continued)

B: Textile/apparel market

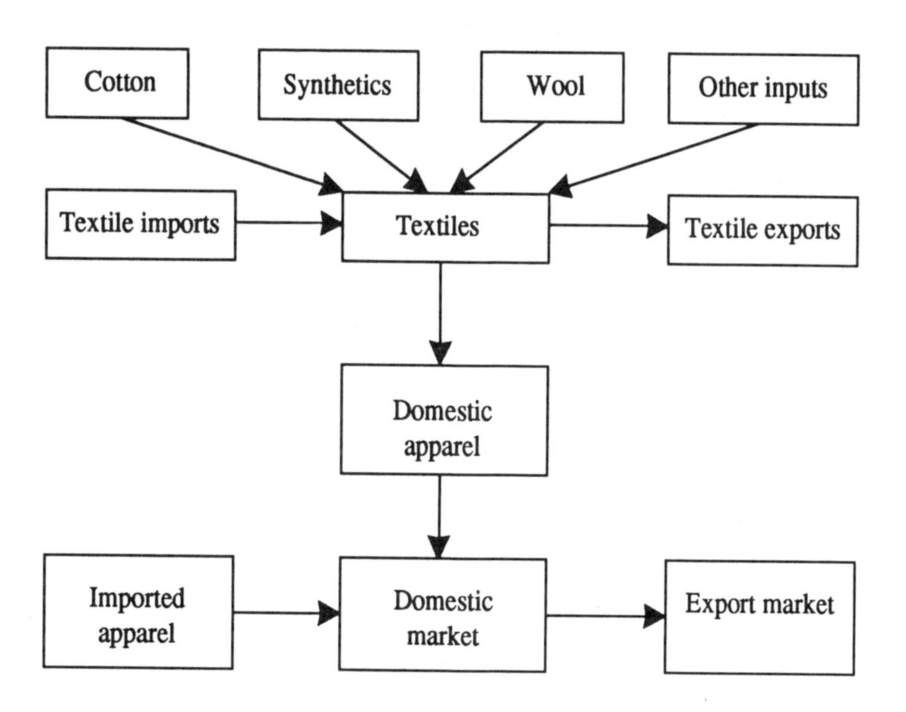

C: World market for apparel

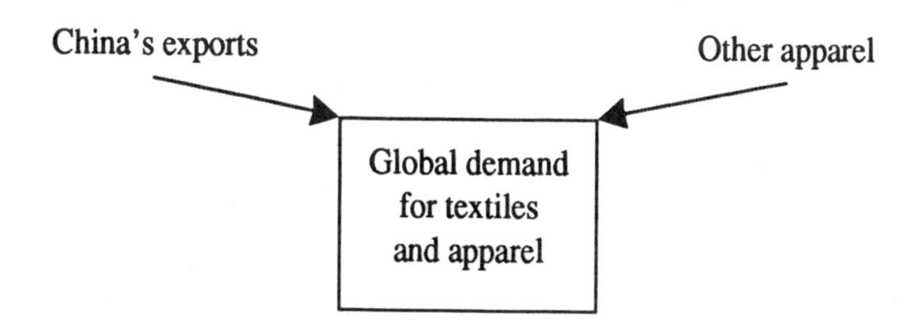

Part C of Figure 4.1 shows the world market for apparel, in which China's exports compete with products produced in other countries. An outward shift in China's export supply schedule has both substitution and expansion effects within the world market. The effect is an increase in China's market share at the expense of other apparel producers. The market expansion effect is the increase in the size of the total market brought about by the increase in the supply of exports from China. Protection in the world apparel market under the Multifibre Arrangement reduces both the substitution and the market expansion effects of increases in China's exports (Martin and Suphachalasai, 1990).

The overall effects of changes in China's foreign exchange policies on the world market for fibres depend upon their impacts both on domestic demand for wool in apparel, and the effects of increased Chinese apparel exports on the global demand for apparel wool. Clearly, any model of the system under study needs to capture these important features of the market.

The effects of the exchange rate system are general equilibrium in nature. The exchange rate affects, and is affected by, all of the key aggregates in the system and any satisfactory model must take into account these interactions. Even the simple graphical model developed in Chapter 3 was general equilibrium in character.

Unfortunately, neither the simple graphical analysis nor the qualitative analysis in the previous chapter can allow us to capture the inter-relationship of immediate concern between macro and microeconomic variables, such as wool imports. The number of variables involved and the complexity of their inter-relationships also rules out the use of purely theoretical models.

One means of overcoming this problem is to develop a computable general equilibrium model of the Chinese economy, with particular disaggregation of the wool, cotton and synthetic fibre industries, and the textile and garment sectors. Dervis, de Melo and Robinson (1981) have demonstrated the feasibility of using such a model to analyse foreign exchange shortages and models of this type can incorporate considerable detail at the individual industry level.

The ease of constructing models of this type has improved dramatically following the development of the GEMPACK suite of programs (Codsi and Pearson, 1988). Two World Bank papers (World Bank, 1985a, 1985b) provide a consistent input-output table for China and estimates of many of the relevant parameters (such as income and price elasticities) needed to construct a model of the ORANI type (Dixon et al., 1982). The major shortcomings of these data are the lack of distinction between domestic and imported intermediates in the intermediate use matrix and the low degree of aggregation of the agricultural sector (into crops and animal husbandry) and the textiles and clothing sector (one industry). The dated nature of the database (1981) is also of some concern given the dramatic changes in the structure of the economy in recent years. More fundamentally, the database is based largely on the official prices, rather than the free-market prices which are relevant for resource allocation at the margin (see Chapter 3).

In recent years, models of the Computable General Equilibrium (CGE) type, with appropriate modifications, have been used to analyse post-reform centrally planned economies (see, for example, Kis, Robinson and Tyson, 1986). In applying this technique to contemporary China, a number of adaptations of the techniques used in modelling other developing countries (Robinson, 1989) were required. The more important of these adaptations were modelling the effects of the foreign exchange retention scheme and adapting the input-output and price data to reflect open-market rather than official prices for material inputs.

Given the major changes in the Chinese economy associated with policy changes and rapid economic growth, any modelling exercise can lead, at best, only to a highly stylised representation of the economy. Even given this constraint, modelling work can provide many useful insights. It provides an explicit framework for analysis, which frequently leads to the discovery of important, but otherwise overlooked, causal linkages. Thus, for instance, Dixon et al. (1982: 61) discovered the 'Keynesian' behaviour of the otherwise neoclassical ORANI model in response to a demand shock after applying such a shock to their numerical model. Similarly Stoeckel's (1979) small CGE model of the Australian economy highlighted the major differences between a resources boom resulting from a mineral discovery and one resulting from an increase in the price of minerals. The objective of this modelling exercise is to develop a model which, while highly stylised, does capture the main economic features involved and provide some insights into the operation of the system.

BROAD FEATURES OF THE MODEL

The model developed in this chapter is of the broad ORANI type (Dixon et al., 1982; Vincent, 1985). Like most other models in this tradition, it focuses on the real side of the economy, with particular emphasis on the response of the economy to trade policy changes. Explicit modelling of most of the nominal income flows in the economy is eschewed, a simplification which seems reasonable for a model focusing on trade policy and which is virtually necessitated by the complex structure of the taxation system in China (Blejer and Szapary, 1989).

The behavioural assumptions of the model involve cost minimisation by producers and utility maximisation by households, and the assumption that there is sufficient competition for unit profits (at the margin) to be driven to zero. The crucial assumption is that economic agents respond to the secondary market prices for inputs and outputs, rather than official prices. The official prices are thus irrelevant to the behaviour of the model, as was demonstrated in the previous chapter, drawing on earlier studies by Byrd (1987) and Wu Jinglian and Zhao Renwei (1987). Although it is recognised that the income distributions induced by divergences between the official and market prices may have an impact on demand behaviour, this second-round effect seems likely to have relatively minor impacts on the results and hence has not been incorporated in the model.

While agents are assumed to respond in a manner consistent with neoclassical theory to the market prices which they experience, these market prices are affected by distortions such as the exchange rate overvaluation, the foreign exchange retention arrangements and import tariffs and licensing, all of which can be incorporated in the model.

Major features which this model shares with others of the ORANI type are that it is linear in percentage changes and that domestic and imported products are imperfect substitutes. Another standard feature of these models is a two level representation of technology in which intermediate inputs are demanded in fixed proportions to the output level of each industry and total primary factors are also demanded in fixed proportions to output. Changes in output levels thus require changes in the level of primary factor inputs which, in the presence of any fixed factor, require substitution between factors. In general, this is represented using Constant Elasticity of Substitution (CES) technology, although the assumption of Leontief technology (fixed proportions) is made for most intermediate inputs.

In the model developed in this chapter, several adaptations to this general pattern are made to account for particular features of the system under study and the available data. Because of the focus of the study on the fibre market, the Leontief or fixed proportions assumptions are too strong in the case of fibres used in the textile industry. Thus this assumption is relaxed to allow for inter-fibre substitution according to a CES technology.

For many goods, there are marked differences between the characteristics of the product produced for the export market and that produced for the domestic market. These differences involve both the physical characteristics of the goods and the package of services involved in their marketing (see Chapter 7). To capture these differences, it is assumed that products sold on the domestic market are differentiated from those sold on the export market. These differences are represented using a Constant Elasticity of Transformation (CET) functional form (Robinson, 1989; Powell and Gruen, 1968).

In the current short-run version of the model, capital is assumed to be fixed in each sector. It would be straightforward technically to build a longer run version of the model in which the capital stocks in each industry were endogenous, although this would introduce some potentially serious difficulties in modelling investment behaviour in China. To simplify matters, and in the absence of a well developed theory of investment for China, investment in each sector has been specified as simply changing in line with total real absorption. As is common in short-run models, investment does not add to the effective capital stock. The underlying time period is assumed to be sufficiently long for new equipment and machinery to be installed but not brought into production.

Given the focus of the study on foreign exchange and trade policies, it seems reasonable to omit an explicit representation of fiscal behaviour. Instead, the model is constrained so that the total real absorption is held constant. Implicitly, it is assumed that the authorities make whatever adjustments to fiscal and monetary policies are needed to achieve this outcome.

While the model focuses on the behaviour of real variables, a skeletal monetary sector is also incorporated. This specification is based on the evidence by Chow (1987) and Feltenstein and Ziba (1987) that the quantity theory of money provides a useful starting point for explaining the price level in China, or at least the behaviour of 'virtual' prices corresponding to those in a market economy.

DATA AND STRUCTURE

The broad structure of the model was determined by the problem under study subject to the availability of data. As discussed above, the initial source of data was the World Bank (1985a: 55-6) table for 1981, which was the latest original table available at the time the model was constructed. The World Bank table also has the advantage of being prepared in a manner consistent with the international SNA conventions, rather than the Material Product System used in Chinese input-output tables. Some additional data was obtained from other sources, including the official input-output table for 1981 (China, State Planning Commission and State Statistical Bureau, 1987).

In the development of standard CGE models, it is assumed that the economy is in equilibrium in the benchmark year. Under this assumption, the model is then 'calibrated' to replicate the benchmark data set if the model is solved in levels. Alternatively, if the model is to be solved in percentage changes (Johansen's method), the full set of parameters is built up from a relatively parsimonious set of parameters estimated or obtained from other sources, together with the equilibrium value shares in the model (see, for example, Dixon et al., 1982).

Clearly, the conventional approach of assuming that the value shares in the model were in equilibrium would not be appropriate given the non-marginal nature of official prices under the two-tier pricing system. To make the model operational, it was assumed that the (largely) planned system operating in 1981 resulted in the same set of quantity variables as would have resulted from an equilibrium market system, albeit one distorted by a wide range of distortions to expand some sectors and cause others to contract. Under this simplifying assumption, a data set corresponding to a market equilibrium could then be obtained by adjusting the price variables. These price adjustments were made with a set of relativities for official and free-market prices collected when a fairly well developed set of secondary markets was operating. The details of the construction of the market price database are given in Appendix 4A.

The price adjustment process changed the gross output value of all industries. Under our assumption that official prices do not affect marginal outcomes, this change in output values must accrue as changes in profits. This assumption seems reasonable in the light of the widely held proposition that the official pricing system leads to major distortions in the relative profitability of different industries in China (see, for example, Chen Xikang, 1988). As a consequence, returns to capital were made the balancing item. The resulting estimates of factor intensities appear to be more consistent with expectations, and with the range of estimates observed in market economies than the set of estimates obtained in the non-adjusted table.

The input-output table on which the model is based is presented in Appendix 4A. It contains 27 sectors: 22 of the 23 sectors presented in the original World Bank table, together with the following sectors of particular relevance to this study—wool production, cotton production, chemical fibre production, textile production, and apparel production.

A prototype version of the model was originally developed with sectors other than those of particular interest compressed into only five sectors. However, this approach obscured some of the important input-output linkages in the model and involved time-consuming data aggregation. Since the use of 27 sectors presented no computational problems with the GEMPACK software, the final model used all of the available sectors.

Separating out the sectors of particular interest (such as wool production, textiles, wool, cotton and chemical fibres) necessitated the use of information from a wide range of sources. Similarly, the elasticity parameters used in the study were obtained by surveying the estimates obtained in other countries for which empirical estimates are available. Details of the model are provided in Appendix 4B.

SIMULATION RESULTS

The model developed in this study was used to examine the impact of a 10 per cent nominal devaluation (increase) of the official exchange rate (yuan/US$). As a simplification, it was assumed that the retention rate for foreign exchange was held constant at 25 per cent in all industries. While the retention rate is known to be substantially higher in some industries, and in some regions, the effective retention rate may be lower in others. Furthermore, the results provide an indication of the effects of liberalisation from the base set of conditions.

The results for the first simulation experiment are presented in Table 4.1. The results in column 1 refer to the model with all parameters set at their base values (see Appendix 4A for details). The results in the second column are based on the modified model, constructed using the 1986 trade structure.

The results presented in Table 4.1 suggest that the effects of exchange rate devaluation on the Chinese economy could be both large and beneficial. Under all of the scenarios considered, a depreciation of the official exchange rate leads to a substantial fall (appreciation) of the secondary market exchange rate; that is, a fall in the cost of foreign exchange. The increase in the supply of foreign exchange generated by the greater incentive to export alleviates the shortage of foreign exchange. This brings about the large consequent fall in the secondary market; that is, an appreciation of the secondary market exchange rate. This result means that the devaluation needed to remove the gap between the official and the secondary market rate would be considerably less than the initial gap between the two variables.

An important consequence of the devaluation is the large expansion it brings about in the size of the traded goods sector. Even a 10 per cent official exchange rate brings about an increase in export volume of 11.4 per cent with the 1981 trade structure in the base case. Total imports also grow substantially. The appreciation in the secondary market lowers the cost of imports leading to a reallocation of demand towards imports and lowering the prices of consumer goods in China. The balance of trade improves noticeably because of the increase in real income, and the improvement in the price of exports relative to imports.

Table 4.1 Simulated effects of a 10 per cent devaluation of the official exchange rate: percentage change from control

	1981 trade structure	1986 trade structure
A. Macro variables		
Market exchange rate	-6.6	-6.1
Export volume	11.4	13.2
Import volume	10.8	8.8
Balance of trade (% of GDP)	0.8	1.2
Real GDP[a]	0.4	0.8
Wage rate	0.3	1.3
Consumption price deflator	-	-
B. Sectoral variables		
Import prices		
Cotton	-6.5	-5.9
Wool	-6.5	-5.7
Textiles	-6.4	-5.9
Apparel	-6.4	-6.0
Import volumes		
Cotton	8.3	12.5
Wool	8.3	9.5
Chemical fibres	6.0	8.8
Textiles	15.6	12.7
Consumer prices		
Textiles	1.1	-0.2
Clothing	0.7	-0.1
Household demand		
Textiles	-0.2	0.1
Apparel	-0.1	0.1
Exports		
Textiles	10.2	13.7
Apparel	11.9	15.5
Output		
Cotton	-1.7	-0.3
Wool	-3.2	-2.6
Textiles	0.7	0.6
Apparel	1.1	3.3

Note: a Measured as the change in nominal GDP deflated by the price of absorption.

Real GDP grows slightly in the base case, with a 10 per cent devaluation resulting in an increase in real GDP of 0.4 per cent. Using the 1986 trade shares, the gains are substantially greater, with GDP rising by 0.8 per cent. These gains measure only the short-run static gains from improved resource reallocation and need to be considered together with the important longer run gains in income which are likely to result from increased openness of the economy. Based on Feder's (1982) analysis, the additional gains from intersectoral externalities associated with the large increases in the size of the export sector considered here are likely to be substantial. These dynamic gains would, in turn, reinforce the effects on growth of GDP, exports and imports considered in this static analysis.

Another important macroeconomic effect of real exchange rate devaluation is its impact on real wages. The expansion of relatively labour-intensive export-oriented activities results in an increased demand for labour and hence in real wage rates. With the original 1981 trade database, this effect was relatively small. However, with the change in the trade structure between 1981 and 1986, this effect became much more important. From the results using the 1986 trade shares, a devaluation would have beneficial effects on employment and/or real wages as well as its effects on the trade sector and on real GDP.

At the sectoral level, the effects of devaluation are particularly interesting. The domestic prices of all imports fall, and the volumes of all of the import categories of particular relevance to this study rise substantially. Using the 1981 trade structure, a 10 per cent devaluation was estimated to result in an increase in wool imports of 8.3 per cent. This result is clearly consistent with the widely expressed view (Angel, Simmons and Coote, 1988) that the shortage of foreign exchange has imposed severe constraints on Chinese wool import demand (see also Chapter 6).

As was noted in the previous chapter, China changed from being a major importer of cotton in 1981 to being a major exporter by 1986. To the extent that wool and cotton are substitutes in the production of textiles, this change might be expected to make the effects of devaluation on the demand for wool more strongly positive. Following this switch in the trade pattern, devaluation would lead to an increase in the price of cotton while reducing the prices of imported wool and synthetic fibres. Thus producers would have an incentive to substitute away from wool towards cotton and synthetics. Consistent with this, the increase in wool imports following a devaluation was larger using the 1986 trade structure, even though the increase in total imports was not as large. Since 1988 China has again become a net importer of cotton (US Department of Agriculture, 1990) and the 1981 trade structure is more relevant while China remains an importer.

Comparison of the results using the 1986 and the 1981 trade structure reveals a number of interesting contrasts. The growth in exports was greater and in imports smaller using the 1986 trade structure. Presumably, this reflects the shift in relative emphasis in exports from oil and other primary products to more labour-intensive and price responsive manufactured exports over the period (Yang Yongzheng and Tyers, 1989: 20). This change results in larger gains in real GDP, and a larger positive effect of real exchange rate on the demand for labour and hence the real wage rate.

At the sectoral level, analysis using the 1986 trade structure yields noticeably different effects on the fibre, textile and clothing sectors. The combination of the favourable effect on fibre prices, together with greater gains from improved export returns, leads to larger output increases in the textile and apparel sectors. Average user prices of wool, cotton, textiles and apparel all decline and household consumption of textiles and clothing rise slightly under this scenario. This result is in contrast to 1981 when consumer prices rose and demand fell. One difference was that in 1981 the penetration of imported intermediate products into the industry was much less. As a result, the change in consumer prices was dominated by the increase in prices of exportable products as a result of the devaluation. In 1986, however, the effect of falling prices of intermediate inputs was more important and the net effect on consumer prices was a slight fall.

An important feature of the simulations is the frequency with which they result in increases in two-way or intra-industry trade. Exports of textiles and apparel expand following a devaluation, but imports of textiles are also projected to expand. At the level of aggregation used in this study, two-way trade is a major feature of the trade structure. Increases in intra-industry trade, like increases in traditional specialised trading activities, allow gains to be obtained from exploitation of comparative advantage in resource availability or skills and expertise.

The overall effects on wool and cotton production in China appear to be slightly negative. Increased output of textiles has a positive effect on the demand for fibres. However, this output is outweighed by the increased competitiveness of imported fibres and consequent substitution of imported fibres for domestic fibres. The adverse effects on domestic output of natural fibre are smaller with the 1986 trade structure partly because of the larger output effect for textiles.

The key conclusion for the wool textile trade to emerge from the simulations using the 1981 and 1986 trade structures is the same. A sustained real devaluation of the official exchange rate would alleviate the foreign exchange constraint and lead to a substantial increase in the demand for wool imports.

While the major source of this import demand would be for re-export, there may be, under some circumstances, an increase in domestic consumption as well. Even where domestic consumption is affected negatively, the magnitude of the adverse effect on domestic demand for textiles and apparel seems likely to be relatively small.

SUMMARY AND CONCLUSIONS

The basic purpose of this chapter was to provide an initial assessment of the likely implications of changes in China's foreign exchange and foreign trade regime on the demand for wool, and hence to develop an approach to analysis of the problem.

The question addressed in the previous chapter was whether the major sectors of the Chinese economy are now sufficiently price responsive for the exchange rate to have a major influence on behaviour. From all the available evidence examined, it seems clear that this is generally the case even though important elements of planning remain and the transmission of price signals is still inhibited in some cases. While these factors will tend to reduce the responsiveness of the economy to price changes, it seems that changes in prices resulting from the foreign exchange system are likely to have major effects on resource allocation throughout the economy.

All the evidence surveyed points to the conclusion that the yuan is currently overvalued, with the effect of discouraging exports and imports and encouraging the production of import substitutes such as many consumer goods. By choking off socially desirable exports and creating a scarcity of foreign exchange, the overvaluation actually makes the price of foreign exchange higher for those who need it than would be the case under a more liberal regime. Consumption of exportable goods is artificially encouraged while consumption of imported goods is generally discouraged. By discouraging socially desirable exports and imports, the overvaluation also lowers national income.

The effects of exchange rate overvaluation on the demand for wool imports were examined in this chapter in the context of the economy-wide effects considered above. The general shortage of foreign exchange resulting from the overvaluation would be expected to reduce the ability to import products such as wool. The high price of foreign exchange associated with its shortage would also be expected to reduce the demand for wool. However, the presence of trade at several levels of the processing chain complicates the analysis of the effects of devaluation. While production of exportable textiles and apparel would undoubtedly increase, the rise in the price received for exports may decrease the domestic demand for these exportable products.

Based on the analysis of the Chinese economy and its fibre, textile and clothing sectors in particular, a 27-sector Computable General Equilibrium model was specified and constructed. The distinguishing feature of this model was its use of market, rather than plan, prices throughout. The model was constructed to represent the behaviour of the post-reform economy in which the important determinants of economic behaviour are the prices received (paid) at the margin. Its 27 sectors were based primarily on those included in the World Bank (1985a) study of the Chinese economy, but with additional attention devoted to the fibre, textile and apparel sectors.

The model was used to explore the implications of a reduction in the overvaluation of the currency brought about by a devaluation of the official exchange rate, with appropriate monetary and fiscal policies operating to control the price level and the level of aggregate economic activity.

Under these circumstances, a devaluation was found to have major beneficial effects on the economy. In the base model, a 10 per cent devaluation of the official exchange rate was found to cause an appreciation of 6.6 per cent in the secondary market exchange rate (in other words, the cost of foreign currency in terms of yuan falls in that market), substantially reducing the cost of all imports.

Devaluation was also found to result in a very substantial increase in the level of exports and imports, with a 10 per cent devaluation resulting in an increase of 11.4 per cent in total export volume. GDP increases slightly following the devaluation and the balance of trade improves. The expansion of relatively labour-intensive export industries also leads to an increase in the demand for labour and hence the wage rate. Devaluation also leads to an increase in intra-industry trade, with both imports and exports from each industry being stimulated.

Devaluation causes a substantial increase in the demand for imported wool, with a 10 per cent depreciation resulting in an increase of 8-10 per cent in wool imports. The increase in demand for wool is primarily the result of substantial increases in exports of textiles and clothing. The effect on domestic consumption levels seems to be relatively small, with a slight decline occurring using the 1981 trade shares and a slight increase when the 1986 trade shares are used. The adverse effect of the devaluation on average consumer prices of domestic textiles and apparel is mitigated by increases in lower priced imports of intermediate products. This effect was larger in 1986, when imports of textiles had increased as a share of total consumption.

Overall, it appears that removal of the current serious overvaluation of the exchange rate would have a number of important advantages for China through its beneficial effects on export performance, level of income and trade balance. Such a policy would also mitigate one of the constraints which has constrained the growth of exports of wool to this important and rapidly developing market.

5 China's industrialisation and fibre self-sufficiency

KYM ANDERSON

INTRODUCTION

The dramatic reforms to China's economic policies since the late 1970s have stimulated extremely rapid growth in this populous economy. Much of the output growth has occurred in the countryside, where higher farm product prices, freer rural markets and the shift from team to household responsibility for management decisions have encouraged rural output to expand substantially. During the first half of the 1980s, food output per capita rose by one-quarter in China compared with a rise of only 3 per cent in the rest of the world, and cotton output nearly doubled between 1980 and 1984. Industrial production in rural China also increased enormously. However, with higher incomes and fewer controls on rural markets, and with rapid growth in textile production for export, the consumption of food, feed and fibre also is increasing rapidly in this still quite poor country. Is China likely to be a substantial net importer or net exporter of agricultural products over the longer term, or will it instead remain close to self-sufficient as in the past? In particular, what will happen to China's trade in fibres? The latter question is important not only for China but also for the rest of the world, and especially for fibre-exporting countries, given that China accounts for one-quarter of the global cotton market and one-eighth of the world's usage of wool.

To begin to address this latter question, the next section of the chapter draws on standard trade and development theory to highlight the key determinants of an economy's changing structure and comparative advantages as it grows. That theory suggests agriculture will dominate a poor economy as its people struggle to feed and cloth themselves, but that as economic development proceeds agriculture's share of production and employment will fall. If that poor economy opens itself to international trade, its exports will tend to be primary products initially but will gradually include more manufactures as the economy develops. Manufactured exports will emerge earlier, and be initially more unskilled labour-intensive, the more densely populated the country.

This theory is strongly supported both by global cross-country evidence and by time series evidence from East Asia's market economies. But how relevant is it for China's large centrally planned economy that until recently had only a very limited exposure to international trade? The third section of this chapter examines the historical record and finds that even during the first three decades of the People's Republic, when heavy-handed planning influenced production and consumption much more than did international or even domestic product or input prices, the intersectoral changes to production and export specialisation were much as theory would lead one to expect. It is true that the policy reforms which brought a sudden improvement in agricultural incentives in the late 1970s and early 1980s boosted cotton self-sufficiency and induced a temporary reversal in the long-term decline in the relative importance of the farm sector, but the downward trends resumed from the mid-1980s.

Will these downward trends in China's agricultural comparative advantage and fibre self-sufficiency continue? What would China be exporting to pay for an increasing volume of agricultural imports? And what policy changes might be introduced as the economy develops — particularly in the light of the macroeconomic problems and political instability of the late 1980s—that would alter the trade pattern which would otherwise evolve? No definitive answers can be given of course, but the weight of the arguments presented in the third section of this chapter suggest agricultural self-sufficiency will continue to decline, at least for 'feedgrains' and fibres. An important qualification to that conclusion, however, has to do with the rest of the world's preparedness to continue to do business with China and in particular to allow it to exploit its comparative advantage in light manufactures such as textiles and clothing in the 1990s and beyond.

DETERMINANTS OF COMPARATIVE ADVANTAGE IN A GROWING ECONOMY

Much of the production and employment of a low-income economy involves the provision of essentials, namely food and fibre. Agriculture's shares of GDP and employment thus start at high levels. However, as economic development and commercialisation proceed, agriculture's relative importance typically falls. This phenomenon is commonly attributed to two facts: the slow rise in the demand for food and fibre as compared with other goods and services as incomes rise, and the rapid development of new technologies for agriculture relative to other sectors which leads to expanding farm output per hectare and per worker (Schultz, 1945, chs 3–5; Kuznets, 1966, ch. 3; Johnson, 1973, ch. 5). Together, these two facts ensure that in a closed economy (including the world as a whole) both the quantity and the price of agricultural relative to other products will decline, as will the share of employment in agriculture. The evidence certainly supports the declining importance of agriculture in world production and employment and even appears to support the view that the long-run trend in agricultural prices relative to industrial product prices during this century has been downward.[1]

But what about an open economy which has the opportunity to trade at the international terms of trade? Consider a small open agrarian economy which could trade all of its products at those declining international terms of trade for agriculture. Its agricultural sector would decline in relative importance unless its own productivity growth is biased towards agriculture sufficiently for the relative output changes to more than offset the adverse change in the terms of trade that result from economic growth abroad. This agricultural bias in productivity growth would have to be even stronger in a *large* open economy because its own contribution to world agricultural exports would depress the terms of trade even further.

Moreover, in reality, many goods and services are such that transaction costs make them prohibitively expensive to trade internationally. Insofar as these non-tradables as a group tend to have a high income elasticity of demand and/or to be produced in industries which have a relatively low rate of labour and total factor productivity growth, the share of non-tradables in output and employment will rise over time. There is strong evidence to suggest that the income elasticity of demand for services is above unity (see, for example, Lluch, Powell and Williams, 1977; Kravis, Heston and Summers, 1983; Theil and Clements, 1987). So that may well be true for non-tradables as a group too, given that the bulk of non-tradables are services. Evidence on the productivity growth in the non-tradables sector in total is difficult to find, but it would not be surprising if it was below that for the tradables sector simply because the 'cold-shower' effect of international competition on the former is absent. For these two reasons the share of tradables in GDP and employment is more likely to decline than to increase in a growing economy.

If there is a tendency for agriculture's relative importance in the tradable part of an open growing economy to decline, and if the tradables part of the total economy is likely to decline, then the combined effects of these tendencies multiply the likelihood of agriculture's relative demise over time. For that not to happen, agricultural productivity growth has to be sufficiently greater than productivity growth in the other sectors so as to offset the effects of the decline in the relative price of agricultural goods.[2]

One might also expect agriculture's share of a country's exports to decline over time, though again this could be avoided by rapid agricultural productivity growth at home. The decline in the relative price of agricultural products in international markets due to the rest of the world's economic growth would, *ceteris paribus*, discourage domestic farm production and encourage domestic food and fibre consumption while doing the opposite in the domestic market for non-farm tradables. Only by exceptionally rapid farm productivity growth in one's own country could this be avoided — and even then for a large country its agricultural growth would have to be sufficiently fast to offset the adverse effects its own expansion would have on further depressing the international terms of trade for agriculture.

Must agricultural self-sufficiency also decline as an economy grows? The modification to standard trade theory provided by Krueger (1977) is helpful in answering this question (see also Deardorff, 1984; Leamer, 1987; Eaton, 1987). It involves in its simplest form a model with two tradable sectors, producing primary products and manufactures, and three factors of production: one, natural resources, which are specific to the primary sector; two, other broadly defined capital which is specific to the manufacturing sector; and three, labour which is used in both sectors, is intersectorally mobile and exhibits diminishing marginal product in each sector. In this model, at a given set of international prices, the real wage rate is determined by the overall per worker endowment of natural resources and capital, while the pattern of comparative advantage between manufactures and primary products is determined by the relative endowments of man-made capital and natural resources.

An underdeveloped country with little capital will produce mostly primary products and export them (in raw or lightly processed form) in exchange for manufactures. As the availability of industrial capital per worker expands, wages increase and labour is attracted to the manufacturing sector. The country gradually changes from being an exporter of predominantly primary products to being an exporter of predominantly (non-resource-based) manufactured goods, with the capital intensity of exported manufactures increasing over time. When labour begins to be attracted to manufacturing at an earlier stage of economic development, and the non-resource-based manufactured goods initially exported use unskilled labour relatively more intensively, the lower the country's natural resources per worker and hence initial wage rate. This is because the low wage will give the country poor in resources an international comparative advantage initially in labour-intensive, standard-technology manufactures (Balassa, 1979; Balassa and Bauwens, 1988).

Relaxing the assumption that capital is not required in addition to natural resources and labour in primary production strengthens the conclusion that densely populated countries poor in natural resources will begin manufacturing at an earlier stage of capital availability per worker than countries rich in resources. Capital is a complement to natural resources at early stages of economic development because it is needed initially to clear land (or develop mine sites). Only after farm land is highly developed does capital become a possible substitute for agricultural land. The greater share a country would tend to employ of its available capital in primary production rather than in manufacturing, the greater its agricultural land and mineral resource endowments per worker.

Also, relaxing the assumption that capital is not internationally mobile allows the possibility of a country proceeding faster along its path of economic growth than its domestic savings rate alone would allow. These changes in comparative advantage can proceed even more rapidly when barriers to foreign capital inflow are lowered.

The demand for food increases with population and per capita income while the demand for fibres (and other industrial raw materials) increases with textile (and other industrial) production. Thus relatively rapid increases in a country's GNP and manufacturing output raise domestic relative to overseas demand for primary products and hasten the country's switch from being a net exporter to being a net importer of primary products.

The country's comparative advantage in agricultural products as compared with minerals and energy products depends largely on the ratio of agricultural land to mineral resources in this country relative to the rest of the world. The higher that ratio relative to the world average, the more likely it is that the country's primary exports are dominated by agricultural products and/or that its primary product imports are dominated by minerals and energy.

To summarise so far, this theory suggests that a poor country opening up to international trade will have large shares of production and employment in the primary sectors, particularly agriculture, but that these will decline with economic growth. Initially, the country's trade will tend to specialise in the export of primary products, though less so the more densely populated the country. The proportion of those primary exports that is agricultural will be larger, the greater this country's endowment of agricultural land relative to minerals compared with that ratio globally. If the country's domestic income and capital to labour ratio grow more rapidly than the rest of the world's, its export specialisation will gradually switch away from primary products (in raw or lightly processed form) to manufactures in the absence of distortionary policies. The manufactured goods initially exported will be more labour-intensive the more poor in resources or densely populated the country.

Since many textile and clothing production activities tend to be intensive in the use of unskilled labour, this theory suggests that they will be among the items initially exported by a newly industrialising, densely populated country, and that as the demands for textile raw materials by that country's expanding textile industry grow, so the country's net exports of natural fibre will diminish, or net imports of natural fibre will increase, other things being equal.

This theory has strong empirical support from both cross-sectional and time series evidence. (See, for example, Kuznets, 1971; Chenery, Robinson and Syrquin, 1986; Anderson, 1987.) The negative relationship between agriculture's shares of gross domestic product, employment and exports on the one hand, and income per capita on the other, are found to be significant statistically, and these shares are also negatively associated with population density per unit of agricultural land (Anderson, 1989).

Two other indicators of changes in comparative advantage are useful for present purposes. One is an index of export trade specialisation, or what Balassa (1965) called 'revealed' comparative advantage, defined as the share of a product group in one country's exports divided by that product group's share of world trade. The export trade specialisation indexes reported in Table 5.1 show the monotonic decline over time in the agricultural comparative advantage of East Asia's market economies, to the point where the share of agricultural products in these economies' exports is now less than half the world average. In the 1950s, by contrast, that share was two and a half times the world average for Korea and Taiwan. Table 5.1 also reveals the very strong growth over time in East Asia's comparative advantage in textiles and clothing, before it was lost to more capital-intensive manufactures. At the peak these products were between five and seven times more important to the exports of these economies than to the world as a whole. Japan lost comparative advantage in these items to its newly industrialising neighbours from the 1960s, which raises the question—addressed below—as to the extent to which these newly industrialised economies will similarly adjust for China as it looks to expand its labour-intensive industrial

Table 5.1 Indexes of 'revealed' comparative advantage[a] in agriculture and in textiles and clothing in East Asia, 1899-1986

		Agriculture	Textiles and clothing
Japan			
	1899	>1.0	1.5
	1913	0.7	2.6
	1929	0.7	2.9
	1937	0.5	4.1
	1954-56	0.4	5.5
	1964-66	0.3	2.7
	1971-73	0.2	1.7
	1976-78	0.2	1.0
	1982-84	0.1	0.7
	1985-86	0.1	0.5
Hong Kong			
	1954-56	na	5.4
	1964-66	na	7.0
	1971-73	na	7.6
	1976-78	na	9.2
	1982-84	na	7.2
	1985-86	na	6.9
Korea			
	1954-56	2.7	na
	1964-66	1.6	4.3
	1971-73	0.7	6.3
	1976-78	0.6	6.6
	1982-84	0.5	4.8
	1985-86	0.3	4.1
Taiwan			
	1954-56	2.6	na
	1964-66	2.1	2.2
	1971-73	0.8	4.8
	1976-78	0.7	5.0
	1982-84	0.6	4.1
	1985-86	0.5	3.3

Notes: na — not available.
a Share of an economy's exports due to these commodities relative to those commodities' share in total world exports, following Balassa (1965). Agriculture is defined as SITC sections 0, 1, 2 (excluding 27, 28) and 4; textiles and clothing include SITC division 65 and 84.

Source: Anderson (1989, Table 2.3).

exports. The fact that these indexes have been falling for these economies during the 1980s suggests they are already making such adjustments.The other useful index of agricultural comparative advantage is the ratio of domestic production to domestic consumption, or the self-sufficiency ratio. Again, this has steadily declined in East Asia's market economies during the past three decades,

Table 5.2 Agricultural self-sufficiency and agricultural protection in East Asia, 1961-86

	Self-sufficiency index (production as a percentage of apparent consumption)				
	Food and feed[a]	Cotton	Wool	Silk	Average nominal rate of protection[b]
Japan					
1961-64	78	0	1	115	68
1965-69	77	0	1	91	87
1970-74	67	0	1	77	110
1975-79	67	0	1	81	147
1980-84	64	0	0	81	151
1985-86	67	0	0	75	
Korea					
1961-64	92	8	0	160	3
1965-69	86	5	0	161	18
1970-74	77	3	0	152	75
1975-79	75	1	0	114	14
1980-84	69	0	0	85	195
1985-86	69	0	0	56	239
Taiwan					
1961-64	120	17	0	101	2
1965-69	115	13	0	103	2
1970-74	97	2	0	98	17
1975-79	85	0	0	77	36
1980-84	77	0	0	54	57
1985-86	77	0	0	29	

Notes: a Weighted averages for grain, meat, dairy products and sugar, with weights based on the value of domestic production at border prices.

b The percentage by which the domestic producer price for grain and meat exceeds the international price at the country's border. The final period shown for Japan and Taiwan is 1980-82, and that for Korea is 1985. These estimates are from Anderson, Hayami and others (1986: 22).

Sources: International Economic Data Bank, *Agricultural Tapes* (based on FAO and USDA sources), Australian National University, Canberra, 1978; Anderson, Hayami and others (1986).

when food and feedstuffs are considered as an aggregate, and is now virtually zero for cotton and wool. As the first column of Table 5.2 shows, Japan's self-sufficiency in foods and feeds fell from 78 per cent in the early 1960s to 64 per cent in the early 1980s. For Korea and Taiwan the declines have been even faster, from 92 to 69 per cent and from 120 to 77 per cent, respectively. Moreover, these declines have occurred despite the dramatic increase in the extent to which farmers in these economies have been protected from import competition (see final column of Table 5.2).

THE EXPERIENCE OF CHINA SINCE 1949

Clearly the standard neo-classical theory of structural change and changing comparative advantage in growing economies is supported strongly by the above empirical evidence.[3] But how well does this theory fit the experience of centrally planned China? If it can be shown that China also has changed in ways predicted by the above theory, then that theory and the past experiences of other countries, and especially of the more advanced economies of East Asia, can be used to provide some guidance as to where China's economy might be going in the years ahead.

For much of the period since the communists took over China in 1949, economic growth of the world's most populous country has been reasonably rapid, albeit from a very low base. According to official statistics, national income measured at constant prices rose at an average annual rate of 6 per cent between 1952 and 1978, or at 4 per cent in per capita terms. During the first decade of the dramatic economic reforms begun in December 1978, however, economic growth was spectacular. Official Chinese estimates show real per capita national income to have risen 7.1 per cent per year during 1978–84 and 9 per cent between 1984 and 1987, while the World Bank's estimate for the 1980-86 period is 9.3 per cent. This is well above any other country's performance during that period and more than twice the estimates for China's per capita income growth during the 1960s and 1970s. Moreover, this performance is even more outstanding when one recalls that this is for virtually a whole continent. If one considers just the more progressive eastern seaboard provinces—the population of which exceeds Japan's—the real per capita growth achievement has been more than 15 per cent per year since 1978. The level of China's per capita income by 1985 is believed to have reached US$500, which is double that of India, close to that of Indonesia and one-quarter that of Malaysia and South Korea.[4]

As well as having a rapidly growing economy, China is very densely populated. Its 22 per cent of the world's population is crowded onto 7.13 per cent of the world's land area. China suffers from having a large proportion of its land area that is not useful for agricultural purposes, but it exploits every hectare of its potentially ploughable land so that its share of global land that is arable is only marginally lower at 7.08 per cent as of 1985, according to the Food and Agriculture Organisation's *Production Yearbook*.

One might expect from the previous section that, being densely populated and growing relatively rapidly, China's economy should be structurally adjusting away from agriculture, transforming its comparative advantage from primary products towards unskilled labour-intensive manufactures, and, in particular, reducing its self-sufficiency in cotton and wool as its textile industry expands. And this is indeed what has happened. Agriculture's contribution to what is defined as net material product (which excludes some services from what the OECD would otherwise call GDP) fell from two-thirds to one-third during the first three decades of the People's Republic, while industry has almost quadrupled its share of national output from what in 1949 was only one-eighth. Agriculture's importance was boosted substantially between 1978 and 1982 following the raising of farm prices, the freeing of rural markets and the introduction of the household responsibility system, but it has since resumed its downward slide. As measured by the World Bank, agriculture's contribution to GDP in 1986, at 31 per cent, was between the average contributions of other low-income countries and of lower middle-income economies, which is about what one would expect for the country's per capita income and size (Anderson, 1989, Figure 3.1).

World Bank data also indicate that in 1965 less than one in five Chinese workers was employed outside agriculture, compared with one in four in India and one in three on average in lower middle-income countries. By the end of 1987, however, this ratio for China had doubled and only 60 per cent of workers remained in agriculture, compared with around 70 per cent on average in other low-income countries and 55 per cent in lower middle-income economies (Anderson, 1989, Table 3.5).

The extent of the decline in the share of agricultural products in China's exports is even more marked. Prior to the 1970s more than half of China's exports came from agriculture. Since 1970, however, that share has declined steadily and by 1987 was only 20 per cent. Indeed, primary products in total have reduced enormously their contribution to China's exports: prior to the 1960s they accounted for more than two-thirds of all exports but in 1987 they contributed only one-third (Table 5.3 and Figure 5.1).

The corollary to agriculture's relative decline is the relatively fast growth in

Figure 5.1 Sectoral shares of commodity exports, and exports as a share of national income, China, 1955/57 to 1987 (per cent, value based)

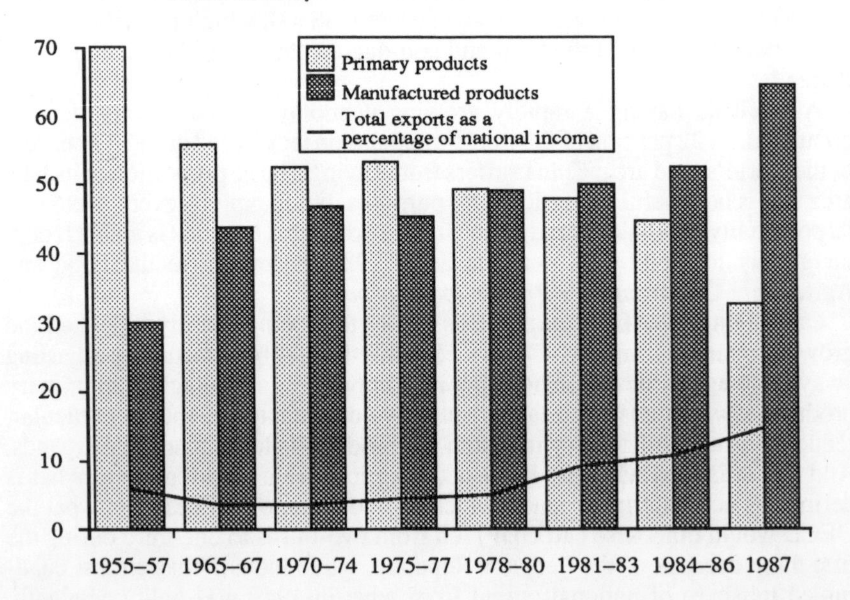

Source: Table 5.3.

Table 5.3 Sectoral shares of commodity exports, and exports as a share of national income, China, 1910-87 (per cent, value based)

	Primary products			Manufactured products			
	Total	Agriculture and processed food	Fuels, minerals and metals	Total	Textiles and clothing	Other manufactures	Total exports as a % of national income
1910-29	85	51	34	15	6	9	na
1955-57	70	55	15	30	14	16	6
1965-69	56	51	5	44	20	24	4
1970-74	53	47	6	47	21	26	4
1975-77	54	38	16	46	21	25	5
1978-80	50	30	20	50	23	27	6
1981-83	49	23	26	51	25	26	8
1984-86	46	22	24	54	29	25	10
1987[a]	34	20	14	66	36	30	14
1978	51	35	16	49	23	26	6
1979	50	30	20	50	24	26	6
1980	51	25	26	49	23	26	7
1981	50	24	26	50	24	26	8
1982	50	23	27	50	24	26	8
1983	49	23	26	51	27	24	7
1984	48	22	26	52	28	24	8
1985	50	21	29	50	26	24	9
1986	38	22	16	62	33	29	11
1987[a]	34	20	14	66	36	30	14

Notes: na — not available.
a Preliminary.

Sources: Anderson (1989, Table 3.7) and, for the 1980s data for the final column, Asian Development Bank, *Asian Development Outlook 1990*, Manila, 1990.

industrial and services employment, and production in mining and manufacturing exports. The share of employment in services has nearly doubled and that in the industrial sector has nearly trebled since 1965, switching China from below to above the average of these shares for low-income economies. The share of industry in GDP is now extraordinarily high, whereas in the early 1950s it was very low by world standards (Anderson, 1989, Table 3.5 and Figure 3.2). This switch from below average to well above average reflects the fact that heavy industry had been favoured while the services sector had been seriously neglected by Chinese planners prior to as compared with since the reforms began in 1978.

The share of manufactured goods in China's commodity exports grew from 35 per cent in the latter 1950s to around 45 per cent in the latter 1960s and early

Table 5.4 Index of 'revealed' comparative advantage[a] in agriculture, other primary products, textiles and clothing, and other manufactures, China, 1955-87

	Agriculture	Fuels, minerals and metals	Textiles and clothing	Other manufactures
1955-59	na	na	2.1	na
1965-69	2.1	0.3	3.3	0.44
1970-74	2.3	0.3	3.4	0.48
1975-77	2.2	0.7	3.9	0.46
1978-80	1.9	0.8	4.6	0.48
1981-83	1.6	1.0	5.0	0.47
1984-86	1.4	1.1	5.1	0.47
1987[b]	1.3	0.7	6.5	0.49

Notes: na — not available.
a Value based share of an economy's exports due to these commodities relative to those commodities' shares in total world exports, following Balassa (1965). Agriculture is defined as SITC sections 0, 1, 2 (excluding 27, 28) and 4; fuels, minerals and metals as SITC section 3 and divisions 27, 28 and 68; textiles and clothing as SITC divisions 65 and 84; and other manufactures as SITC sections 5 to 9 (excluding 65, 88 and 84.)
b Preliminary.

Source: International Economic Data Bank, *Trade Data Tapes* (based on United Nations statistics), Australian National University, Canberra, 1989.

1970s, but then grew even faster following the economic reforms of the past decade. By 1987 two-thirds of exports were manufactured goods (see Figure 5.1). This occurred despite the rapid increase in the value of energy product exports: the share of fuels, minerals and metals (mainly petroleum) in total exports jumped from 6 per cent just prior to the first oil price shock of 1973 to a peak of 29 per cent in 1985, before dropping back substantially in 1986 and 1987 (Table 5.3).[5]

This steady decline in agriculture's share of exports at the expense of petroleum and light manufactures occurred during a period when the value of these products in world trade also was changing. A better index for examining changes in export specialisation is therefore Balassa's (1965) 'revealed' comparative advantage index, which is the ratio of China's sectoral share of exports to that sector's share of total world exports. Estimates for this index are shown in Table 5.4. The decline in China's comparative advantage in agricultural products by this measure is still very strong, as is the strengthening of its comparative advantage in textiles and clothing: between the early 1970s and 1987 the index almost halved for agriculture and almost doubled for textiles and clothing. During the same period this index for fuels, among other products, grew much less rapidly than those products' share of China's exports, however, reflecting the fact that much of that share growth was simply due to the hikes in world oil prices in 1973 and 1979–80.

Even this export specialisation index ignores an important part of the sectoral contributions to foreign exchange earnings; namely, the replacement of imports by domestic production. It is helpful therefore to also examine net agricultural exports (gross exports minus gross imports) as a percentage of total exports. In the latter 1960s and early 1970s agriculture's net contribution to total export earnings was about one-fifth. This fell to one-eighth in the mid-1970s and to zero during the period from 1978 to 1984. It rose briefly to 8 and then 11 per cent in 1985 and 1986 as exports of farm products expanded and imports were curtailed, but was down to less than 6 per cent again in 1987, well below the average for the 1960s and 1970s (Table 5.5).

Table 5.5 Net agricultural exports[a] as a share of total exports, China, 1966-87 (per cent, value based)

1966-69	20.9
1970-73	20.8
1974-77	13.0
1978-81	0.0
1982-84	0.1
1985	7.7
1986	11.2
1987	5.9

Note: a Agricultural exports minus agricultural imports, where agriculture is defined as SITC sections 0, 1, 2 (excluding 27 and 28) and 4.

Source: International Economic Data Bank, *Trade Data Tapes* (based on United Nations Statistics), Australian National University, Canberra, 1989.

This net export indicator of changing comparative advantage shows even sharper changes for the natural fibres and textile and clothing subsectors of China's economy. China's initial comparative advantage in primary products manifested itself in part with the exporting of natural fibres, particularly silk. In the late nineteenth century net exports of fibres accounted for about one-third of China's total exports. Even in the first third of this century they contributed close to one-quarter of total exports. By the 1930s their contribution had fallen below 5 per cent, and then from the 1950s China has been mostly a net importer of fibres (Table 5.6).[6] The reason for the decline is not only the shift in productive competitiveness away from agriculture but also the growth in domestic demand for fibres by China's textile and clothing industries, whose international competitiveness has been strengthening steadily. In the late 1800s and early 1900s textiles and clothing were net import items, absorbing more than one-third of China's foreign exchange earnings, but by the 1930s the country was almost self-sufficient in these products in aggregate. Since then textiles have been expanding their net contribution to export earnings, which by 1986 had reached 25 per cent (Table 5.6).

Table 5.6 Net exports of natural fibres and textiles and clothing as shares of total exports of China, 1874–1986 (per cent, value based)

	Natural fibres	Textiles and clothing
1874-99	30.1	-38.8
1900-29	23.3	-35.3
1930-41	3.9	-6.2
1955-59	-10.8	19.4
1965-69	-1.5	17.8
1975-79	-2.3	19.6
1980	-7.3	17.8
1981	-6.5	14.5
1982	-2.9	16.6
1983	-0.0	20.6
1984	1.2	21.3
1985	0.5	20.6
1986	0.4	25.1

Source: Anderson (1989, Table 3.10).

Table 5.7 Self-sufficiency[a] in agricultural products, China, 1961–86

	Wheat	Coarse grain	Rice	Beef & sheep meat	Pork & poultry	Dairy prod.	Sugar	Sub-total[b]	Cotton	Wool
1961-64	80	97	104	100	101	100	48	98	92	92
1965-69	86	101	103	101	102	100	78	100	97	106
1970-74	93	104	105	101	102	100	79	102	95	98
1975-79	96	103	104	101	101	99	72	101	83	96
1980-84	86	100	101	103	101	99	76	98	111[c]	77
1985-86	87	96	100	104	101	98	80	96	88[c]	53

Notes: a Production as a percentage of apparent consumption (production plus imports minus exports, minus the change in stocks).
b Weighted average based on domestic production of the food products shown valued at 1980-82 border prices.
c The changes in cotton stocks in the 1980s were substantial; on average for the 1980-86 period China was just a little over 100 per cent self-sufficient in cotton.

Source: International Economic Data Bank, *Agricultural Tapes* (based on FAO and USDA sources), Australian National University, Canberra, 1989.

The long-run decline in agriculture's contribution to China's earnings of foreign exchange has not altered the fact that the country still remains largely self-sufficient in food staples, but it has meant that self-sufficiency ratios have been declining slightly since the 1970s. Self-sufficiency in grain (the average for wheat, coarse grain and rice) has fallen from 102 per cent in 1970–74 to 94 per cent in 1985–86, for example. Part of that decline has been because of the increasing proportion of livestock that are being grain-fed. That trend has enabled China to maintain a slight export surplus of meat but has been insufficient to prevent self-sufficiency declining a little in dairy products and substantially in wool. The only major farm products for which production growth has been clearly faster than consumption growth are sugar since the mid-1970s and cotton during the early 1980s (Table 5.7).

While these changes in self-sufficiency might seem minor, it needs to be remembered that China accounts for huge shares of world production and consumption of many agricultural products. In the mid-1980s China consumed 20 per cent of the world's grain and natural fibres and 14 per cent of its meat, up from 16, 13 and 8 per cent, respectively, during 1975-79. Because these shares are so large, changes in China's self-sufficiency ratios can have major effects on international food, feed and fibre markets. Between the early 1970s and early 1980s, for example, China's net imports as a share of world trade rose from 9 to 12 per cent for wheat, from 0 to 5 per cent for wool (14 per cent in 1986) and from 3 to 7 per cent for sugar, while its net export shares fell for rice from 28 to 7 per cent and for non-ruminant meat (pork and poultry) from 29 to 11 per cent (Table 5.8).

The changes outlined above in the structure of China's economy away from agriculture and the decline in China's export specialisation in agriculture in favour of light manufactures is precisely what theory and the experience of other East Asian economies lead one to expect, given China's relatively rapid economic growth and low per capita endowment of farm land. It is true that policies distorted the patterns of change in a way that accelerated agriculture's relative decline prior to 1978, but that also has been the case in many other developing countries.[7] The dramatic reforms to rural policies from late 1978 temporarily reversed the decline in agriculture's contribution to national product, but since 1985 the downward trend has continued. The declining contribution of agricultural products to exports as well as other indicators of China's comparative advantage all point to the agricultural competitiveness of China being on a downward long-run trend, with the early and mid-1980s being but a temporary respite.

Will this downward trend in agricultural comparative advantage continue? How will commodities within the agricultural sector fare? Will China allow itself to become more dependent on imports of farm products, including fibres? What role could textiles and clothing play in China's attempt to pay for a larger volume of agricultural imports? And what will be the implications for international markets of these likely developments? These are among the questions addressed in the remainder of this chapter.

CHINA'S PROSPECTIVE GROWTH AND STRUCTURAL CHANGES

An assessment of the likely direction and speed of China's changing comparative advantages needs to begin by asking: will China's economy continue to grow faster than the rest of the world and remain open to foreign trade in goods, capital and technology? Notwithstanding the economic and political instability of

Table 5.8 China's share of world production, consumption and trade in various agricultural products, 1961–86 (per cent)

	Wheat	Coarse grain	Rice	Beef & sheep meat	Pork & poultry	Dairy prod.	Sugar	Sub-total[b]	Cotton	Wool
Production										
1961-64	7	8	31	0.5	10	0.6	1	9	9	4
1965-69	9	8	37	0.7	15	0.5	2	11	16	4
1970-74	10	8	39	0.9	14	0.6	2	12	15	5
1975-79	12	10	38	1.0	15	0.6	3	12	15	5
1980-84	15	11	39	1.5	18	0.8	4	15	23	7
1985-86	17	10	38	2.0	21	1.0	5	16	22	6
Consumption										
1961-64	9	8	31	0.5	9	0.6	3	9	9	4
1965-69	11	8	37	0.7	14	0.5	3	11	16	4
1970-74	11	8	38	0.9	14	0.6	3	12	15	4
1975-79	14	10	38	1.0	15	0.6	4	12	17	5
1980-84	18	11	39	1.5	18	0.8	5	15	21	8
1985-86	20	11	40	1.9	21	1.0	7	17	24	11
Net imports[a]										
1961-64	11	2	-16	-0.0	-18	0.0	7	6	3	1
1965-69	11	-0	-23	-0.2	-41	-0.1	4	2	3	-1
1970-74	9	1	-28	-0.2	-29	-0.1	3	2	6	0
1975-79	10	1	-13	-0.3	-12	0.2	6	4	8	1
1980-84	12	-0	-7	-0.8	-11	0.2	7	6	10	5
1985-86	9	-6	-7	-1.5	-11	0.7	8	2	-9	14

Notes: a China's imports minus China's exports divided by world exports: changes in stocks are therefore excluded from this table.
b Weighted average using 1980–82 border prices as weights.
Source: International Economic Data Bank, *Agricultural Tapes* (based on FAO and USDA sources), Australian National University, Canberra, 1989.

1988–89, there are good reasons to answer in the affirmative. Some of the issues involved are reviewed in the introduction to this volume. The purpose of this chapter is to examine the likely structural changes that would be associated with further rapid growth in China.

Should the economy continue to grow rapidly, agriculture is highly likely to continue its relative decline as a contributor to GDP and employment while labour-intensive manufactures and non-tradable services expand in importance. Anderson (1989) suggests that by the year 2000 farms will account for probably less than one-quarter of GDP and half of all jobs in China, compared with 30 and 60 per cent, respectively, in 1987. This expected 'de-agriculturalisation' process does not necessarily mean that the countryside would be depopulated and the large cities become sprawling mega-centres. The more likely alternative is that many of the industrial activities will be located in rural areas within commuting distance of existing farm households. In fact, the growth in non-farm employment in the countryside already has been much faster than urban employment growth: during the ten years to 1987, the proportion of Chinese workers located outside urban areas increased from 26 to 39 per cent (Anderson, 1989, Table 3.19).

The land constraint in China is likely to ensure that food and feedgrain production growth will not be able to keep pace with the rapid growth in consumption of these products. The reason China has been able to feed its 22 per cent of the world's population using only 7 per cent of the world's arable land has been not only because its grain yields per hectare are very high by world standards but also because a high proportion of its grain has been consumed directly rather than via the feeding of animals. Prior to 1980 less than one-sixth of China's grain had been consumed by livestock. However, this proportion rose to one-quarter by the mid-1980s, suggesting China is on the way towards the 40-50 per cent level of East Asia's market economies. Even though milk consumption went up 50 per cent and meat consumption doubled per person during the 1978–1986 period, a comparison with Taiwan, South Korea and Hong Kong in earlier decades suggests China's per capita consumption of these products (and of fruit and vegetables) may go up a further 50 to 100 per cent by the turn of the century.

The rise in animal products consumption in China will add to pressures for higher grain consumption in China, in order to feed the livestock. These changes in consumption will therefore force up feedgrain consumption (Anderson, 1989, ch. 4). Chinese policymakers are concerned with overall grain self-sufficiency, that is, in both foodgrains and feedgrains. The growth of demand for direct and indirect human consumption therefore exaggerates the self-sufficiency problem. The Chinese policymakers have, at least in principle, a couple of options.

One is to raise prices of domestic food grain and animal products. This would encourage production and reduce demand for grain. But food is still a major component of urban household budgets and attempts to raise food prices have met with great political resistance. Any attempt to keep consumer prices low while raising producer prices is constrained by the growth in subsidies that would then be required.

Instead, the Chinese policymakers may have to specialise in types of grain production. One option is to permit relatively more rapid growth of feedgrain imports than foodgrain imports. Domestic production would then be specialised in foodgrain (such as rice and wheat) while the import market would be used to supply feed to the expanding domestic livestock sector. This strategy will slow the decline in self-sufficiency in foodgrain production. The use of intensive production methods in the livestock sector, which this strategy implies, may also be consistent with the relatively low arable land area per head of population in China. In other words, this type of livestock production may match China's comparative advantage. There is also evidence of relatively low yields at present in pork production, suggesting scope for productivity gains (Brummitt, 1989).

The choice of strategies in the grain and livestock sector also have important implications for fibre production. Chinese policymakers will be loathe to allow cotton acreage to expand greatly at the expense of grain. Some yield increases may be possible if the cotton/fertiliser price ratio were to be raised and/or research and extension efforts were able to improve existing production technologies, but China's cotton yields are already quite high by world standards. Hence the doubling of domestic cotton production during 1980–84, when production incentives altered dramatically, was an event unlikely to be repeated in the forseeable future. China's cotton production during 1985–87 averaged less than 70 per cent of the record 1984 crop, for example.

Nor does China have a great deal of scope for expanding wool production (see Chapter 8). In the agricultural sectors of China, increasing sheep numbers is likely to be limited by the lack of suitable terrain. Sheep numbers will also be limited by more profitable uses of land, and of feed, for reasons just discussed. That is, the relatively more productive use of feed will be for livestock more suited to intensive production, such as pigs, poultry and even cattle. In the pastoral areas, the fragile environment along the north and northwest borders of the country is already heavily grazed by sheep, to the point where higher stocking rates could do irreparable damage to the ecology and thereby to future grazing potential. This, together with the extraordinarily harsh climate of the region, ensures that pasture management and sheep husbandry research in China is likely to have a much lower expected social rate of return than other agricultural research investments China might contemplate. The most likely source of increase in domestic wool production will come from improving the yield of clean wool per sheep through breed adaptation. Whether it will pay graziers to move in that direction as distinct from improving the meat yield of their animals will depend largely on trends in the wool/meat price ratio facing producers and in particular on the extent of the price premia paid for higher-quality wool (which in the past have been all but non-existent).[8]

Even if China's production of fibres does continue to grow, domestic demand for those fibres is likely to grow even faster. One obvious reason is that with increasing incomes and reductions in clothing conformity in China, it is inevitable that the domestic consumer demand curves for fabrics and clothing will shift out. But, in addition, since the textile and clothing industries are leading China's export-oriented industrialisation drive, the domestic demand for cotton and wool from manufacturing firms producing for export is also growing rapidly. In this respect as well, China is following the experiences of Hong Kong, Korea and Taiwan.[9]

As shown in Table 2.1, Korea in the early 1960s used about the same quantities of natural fibres as China in the early 1970s: just under 3kg of cotton and 0.15kg of wool per year. By the early 1970s Korea's consumption had risen to more than 4kg of cotton and 0.3kg of wool, as had China's by the mid-1980s. During the 1970s, however, Korea's consumption of both fibres more than doubled again. This, together with similar experiences in Taiwan a few years earlier, suggests it would not be unreasonable to expect China's per capita consumption of each of these fibres to double before the year 2000. When converted from per capita to total production, that could represent potential increases of up towards 6 million tonnes of cotton and 400,000 tonnes of wool per year. It could increase even more if China chooses not to allow a rapid expansion of production or imports of substitutes in the form of synthetic fibres, on which much of Korea's and Taiwan's expansion in total fibre use has been dependent.[10] With little scope for expanding domestic production of natural fibres other than by reducing food production, it therefore seems highly likely that China's self-sufficiency in cotton and wool will fall steadily during the decades ahead. For example, if domestic fibre production were to increase by only one-third, say, during the period from 1986 to 2000, and the government were to allow the rest to be imported, then China's net imports of cotton would be more than 5 million tonnes and its net imports of wool would be 500,000 tonnes per year by the turn of the century. This would represent close to a doubling of world trade in cotton and a one-third increase in world trade in wool over mid-1980s levels in gross terms.[11] Such an increase for cotton would be unlikely to eventuate unless set-aside land in the United States could be released to expand production for export, because otherwise a rise in the international price of cotton would accelerate the substitution by textile manufacturers towards more synthetic fibre use. It may be less difficult for the wool market to accommodate an increase of one-third in international demand, however, because of the flexibility of farm production in the major wool-exporting countries.[12]

POLICY IMPLICATIONS AND CONCLUSIONS

The policy implications of this analysis for China are clear. First, if present policies remain unchanged, it is likely China's self-sufficiency in agricultural products, particularly in animal feedstuffs and natural fibres, will decline during the 1990s and beyond. More cost-reducing agricultural research could slow this trend, but it is unlikely to reverse it. This trend is to be expected in a densely populated, rapidly industrialising economy, and certainly should not be interpreted as a failure of China's rural development strategy. On the contrary, it may well be judged by history as signifying a highly successful overall strategy for economic development within which rural development policies played an important part in facilitating rather than trying to frustrate the inevitable long-run process of structural adjustment in the countryside.

Second, domestic prices for agricultural products could be raised to encourage more domestic production and less domestic consumption and thereby reduce the decline in agricultural self-sufficiency. Attempts to raise domestic consumer prices for food staples in recent years have met with severe political resistance in China as in other poor countries, so until food becomes a less important component of urban household expenditure this option will be difficult to implement (Anderson, 1988). And while consumer prices remain low, it is impossible to raise government procurement prices of food products without raising budget outlays to fund the gap between procurement and retail prices.

In summary, China may well follow its East Asian neighbours in becoming more import-dependent on feedgrains and fibres, but in so doing it will be able to slow the decline in foodgrain self-sufficiency.

The strategy of greater feedgrain and fibre import-dependence will be feasible in terms of its foreign exchange cost so long as the rest of the world is prepared to do business with China. The more signals other countries can send China to assure it that relying on grain and fibre imports would not leave the country vulnerable to political pressures from outside, the less wary China will be to go down that route of continuing to remain open and exploiting its changing comparative advantages. In other words, there are important connections between China's domestic policy choices, in relation to reforms of pricing and investment decision-making, and the domestic perception of China's impacts on world markets. The progress of the domestic reforms and their contributions to growth therefore depends a great deal on China's relations with its trading partners. The important relations are not only those with agricultural exporting countries but also with countries that are major importers of textiles, clothing and other light manufactures, because without export markets for the latter China would not have the foreign exchange earnings necessary to import foods or feedstuffs or the need to import fibres.[13]

In short, advanced industrial countries have much to gain by adjusting their economies and policies to developments in China and from assuring China of their desire to strengthen economic ties with that populous nation. Industrial countries that are agricultural-exporting countries stand to benefit directly from any increases in China's net imports of food, feed and fibre. But other industrial countries also will benefit from the growth in China's imports of capital-intensive and high technology-intensive goods that will necessarily accompany industrial development in China. The full extent of these gains to industrial countries will only be realised, however, if those countries do not restrict either the access of Chinese manufactured goods into their domestic markets or sales of farm products to China.

6 Chinese wool textile industry growth and the demand for raw wool

CHRISTOPHER FINDLAY AND LI ZE

INTRODUCTION

The demand for wool is derived from the demand for fibres by the wool textile industry. The rapid growth of that industry in China has contributed to increased wool consumption in China, and to an increase in demand, which has spilled over onto world markets. The aim of this chapter is to identify those factors influencing the growth of the wool textile industry.

In Chapter 2 it was argued that economic growth in China will be associated with rising absolute values of clothing purchases, and possibly even more rapid increases in purchases of wool products. However, in an open economy, using domestic demand is neither necessary nor sufficient to cause the growth of the industry producing those goods. More important is the competitiveness of the industry compared to others.

Competitiveness in turn, as argued in this chapter, will depend on the intensity with which labour is used in this industry, especially compared to other parts of the textile industry. In this chapter, while the intensity of labour use in this sector is compared with the rest of the textile industry, it is also stressed that, even within the wool textile industry, there is considerable variation in labour intensity between the subsectors.

Not only do factor intensities vary, so too do the types of wool consumed by various subsectors. In that case, the performance of the components of the industry will vary according to their access to raw materials of suitable quality. In a free-trade regime, raw material supply constraints need not be an issue for a processing industry like this one. However, in the Chinese case, there are constraints on imports of raw materials which will influence observed performance.

The wool textile sector has grown rapidly but in recent years has suffered from declining capacity utilisation and falling profitability. The review of the factors affecting the industry's performance since 1978 is used to comment on these recent events and to identify the elements of a strategy to tackle the issues associated with those aspects of poor performance.

RAPID GROWTH OF THE CHINESE WOOL TEXTILE INDUSTRY

Characteristics of the industry

Table 6.1 shows the value of output of the textile industry since 1980. Total output value of the industry, measured in terms of constant 1981 prices, grew by nearly 10 per cent per year in the period 1980 to 1987, compared to 13 per cent for the whole manufacturing sector over the same period.

The wool textile sector was the fastest growing in the industry over this period (see Figure 6.1), with an annual average growth rate of nearly 24 per cent,

Table 6.1 Output value, and growth rates[a] of the textile industry by sector[b], China, 1980–87 (million yuan, per cent)

Sector	1980		1981		1982		1983	
Total output value	73546	100.00	85602	100.00	86685	100.00	95604	100.00
Growth %	0.00		16.39		1.27		10.29	
Cotton textile	44947	61.10	51791	60.50	49 531	57.14	53984	56.47
Growth %	0.00		15.23		-4.36		9.00	
Wool textile	3790	5.15	4777	5.58	5985	6.80	6231	6.52
Growth %	0.00		26.00		23.40		5.70	
Chemical fibre	5349	7.27	5346	6.25	5765	6.65	6533	6.83
Growth %	0.00		-3.00		7.84		13.32	

Sector	1984		1985		1986		1987		Av. annual growth rate 1980-87(%)
Total output value	108294	100.00	115540	100.00	123072	100.00	141066	100.00	
Growth %	13.27		6.69		6.52		14.62		9.75
Cotton textile	57688	53.27	65746	56.90	69754	56.68	75452	53.48	
Growth %	6.86		13.96		6.10		8.20		7.68
Wool textile	7320	6.76	9633	8.40	11514	9.36	16495	11.69	
Growth %	17.48		31.60		19.53		43.30		23.38
Chemical fibre	9055	8.36	10330	8.94	12059	9.80	12982	9.20	
Growth %	38.60		14.08		16.74		7.65		13.50

Notes: a All growth rates refer to the rate over the previous year.
b Excludes sectors such as linen textiles, silk and cotton knitted goods.

Source: *China Statistical Yearbook*, 1981: 209, 216; 1983: 223-4, 232; 1984: 202-4; 1985: 315, 317, 325; 1987: 260-1; 1988: 316.

Figure 6.1 Output value of the textile sector, China, 1980–87 (billion yuan)

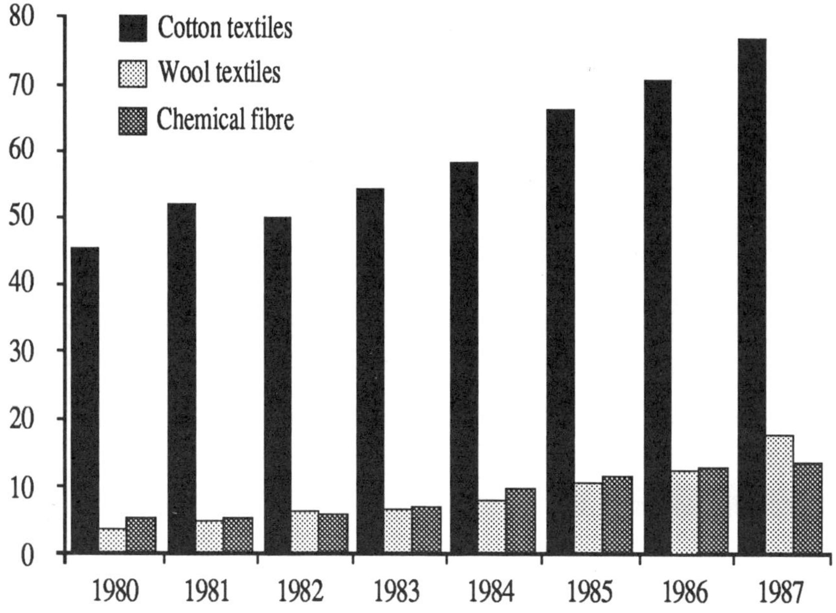

Source: *China Statistical Yearbook*, 1981–1988.

compared to 8 per cent for cotton textiles and 13 per cent for chemical fibre textiles. As a result, the share of the wool sector in the total textile industry has risen from 5 per cent in 1980 to 12 per cent in 1987. The cotton share has fallen from 61 to 54 per cent, while the chemical fibre share has risen slightly from 7 to 9 per cent.

The Chinese wool textile sector accounts for a significant share of world capacity. The world total number of wool spindles is 15.5 million (1986), of which Italy accounted for 3.8m, Japan 2.14m and China 1.99m. China is ranked third in the world and accounts for nearly 19 per cent of the world total of worsted spindles and 13 per cent of woollen spindles.

Labour intensity

The rapid growth of the textile industry as a whole is not surprising. Generally, the industry is regarded as being labour-intensive, and at early stages of industrialisation countries tend to specialise in production of labour-intensive products. This is more likely to occur when the endowment of natural resources is small compared to labour (see Chapter 5), a condition which applies to China relative to its trading partners.

Various indicators can be used to measure the labour intensity of the textile industry and its subsectors. These include labour requirements per unit of output

Table 6.2 Textile industry[a], technology and profitability indicators, China, 1980-87

Item	1980	1981	1982	1983	1984	1985	1986	1987
Textile industry								
Labour cost[b]/value added (%)	12.90	12.70	15.37	16.60	27.89	30.30	32.20	35.50
Value added/output value (%)	29.20	28.60	25.20	21.80	20.80	21.80	23.08	22.80
Valued added/worker[c]	6 949	7 031	5 885	5 527	3 932	4 175	4 268	4 398
Profit & tax/spindle								
Profit & tax/output value (%)	25.68	23.24	19.15	16.05	13.85	14.36	12.41	11.53
Cotton textile industry								
Labour cost[b]/value added (%)	11.70	11.90	14.90	17.50	28.70	32.10	34.16	37.18
Value added/output value (%)	29.50	28.90	24.80	20.50	19.00	20.80	21.40	22.28
Valued added/worker[c]	7 640	7 533	5 986	5 251	3 825	2 944	4 025	4 203
Profit & Tax/spindle	627	6 522	472	369	324	349	378	388
Profit & tax/output value (%)	24.82	23.90	19.23	14.63	12.48	12.22	13.00	13.31
Chemical fibre industry								
Labour cost[b]/value added (%)	8.50	9.60	9.70	9.90	9.70	9.00	9.48	11.60
Value added/output value (%)	25.60	28.70	27.90	27.70	27.50	26.60	26.90	30.90
Valued added/worker[c]	10 514	9 269	9 323	9 209	11 264	14 088	14 490	13 469
Profit & tax/spindle	2 392	2 347	1 971	2 007	2 597	2 234	2 146	2 426
Profit & tax/output value (%)	23.29	25.87	23.33	23.65	23.52	22.46	21.23	24.00
Wool textile industry								
Labour cost[b]/value added (%)	9.70	9.90	10.30	10.80	16.80	16.60	18.20	20.86
Value added/output value (%)	39.90	35.40	35.40	35.50	33.50	35.80	35.60	29.07
Valued added/worker[c]	9 259	9 052	8 780	8 492	6 516	7 635	7 883	7 492
Profit & tax/spindle	2 132	1 875	1 854	1 582	1 475	1 881	1 661	1 524
Profit & tax/output value (%)	33.77	29.22	27.96	25.52	24.29	27.24	24.30	18.40

Notes: a All the data are national data including township enterprises from 1984-87. The unit of valued added/worker is yuan/person. Profit and tax/spindle is yuan/spindle. Other textile sectors such as cotton knitted goods, silk and jute textile, and others are not listed in the table.
b Labour costs include staff and non-productive workers' wages.
c Workers include only productive workers.

Sources: *China Statistical Yearbook*, 1987: 225, 260, 263-6, 321; 1986: 182, 192, 201, 230-1; 1985: 220, 237, 331, 378, 380, 383; 1984: 115, 202, 217; 1983: 129, 224, 234, 298; 1981: 110, 204, 209, 261, 266-7; 1988: 303, 323, 325; *Almanac of China's Economy*, 1985: VI-44; *China Textile Industry Yearbook*, 1984-85: 383.

across industries or labour cost relative to value added. Only data for the value added based indicators are available in this instance.

Table 6.2 shows a number indicators of textile industry performance in China and, of particular note at this stage, the share of labour remuneration in value added and value added per worker. These two variables can be used as indicators of labour intensity although they have different interpretations. The latter measures labour input in terms of availability of labour time (provided by the number of workers) so that capital inputs include not only physical capital but also returns to human skill. The use of the share of labour in value added implies a narrower definition of capital and one which excludes human skill.

Table 6.2 shows that the labour share in value added in 1987 for the whole textile industry was about 35 per cent. The cotton textile sector according to this indicator is the most labour-intensive, with a labour share of value added of 37 per cent compared to 21 per cent for the wool sector and 12 per cent for chemical fibres.

These data refer to 1987, and there has been significant change in the labour cost in value added since 1980. The ratio has nearly trebled in the industry as a whole, and doubled in the wool textile sector. These changes reflect in part the rise in real wages in China since 1978 when the rate of industrial growth increased. The rising real wages had less of an impact in the synthetic sector which is much less labour-intensive. The differential rates of increase of this ratio between the sectors also suggests that the substitutability of labour for capital is much less in the cotton sector and the wool sector compared to the synthetic sector.

The second indicator, of value added per worker, yields similar results. The value added per worker in the cotton sector is about half that in the wool sector, which in turn is about half that in synthetics.

Value added per worker has also changed over time. It would be expected that in the presence of rising real wages, and an incentive to substitute away from labour, labour productivity would rise and that, as a result, value added per worker would rise. This occurred in the synthetic sector. But in the cotton and wool sectors value added per worker fell. This result suggests a qualification to the earlier argument that the rising share of labour costs in value added reflects rising real wages. As we discuss in detail below, capacity utilisation has also been falling, which, in the presence of a fixed labour force, could also contribute to these results.

So far, we have discussed the factor intensity in the wool textile sector as a whole, compared to cotton and synthetics. There is, however, considerable variation in factor intensity even within the wool textile sector.

Table 6.3 and Figure 6.2 show the composition of sales (current price) value of the wool textile industry. By 1987 the largest subsector was knitting wool yarn (38 per cent of the industry's sales value), followed by the worsted sector (22 per cent), then the woollen fabrics and woollen blanket sectors (with 19.8 per cent and 11.2 per cent, respectively), and finally the woollen knitted good sector (8.9 per cent). These sectors make up about 85 per cent of the output value of the wool textile industry. Remaining sectors not shown in the table include top making (that is, the scouring of wool and its preparation for spinning), velvet fabric, industrial woollen fabrics, industrial felt and woollen decoration fabrics.

Table 6.4 includes details of labour share of value added and value added per worker by subsector of the industry. Within the sector, the ranking is in terms of decreasing labour intensity: woollen fabrics, woollen blanket/worsted fabrics,

Table 6.3 **Value of wool textile industry domestic sales[a] and exports, China, 1978–87 (million yuan)**

Sector	1978	1979	1980	1981	1982	1983	1984	1985	1986	1987	1978–87 average annual growth %
Worsted fabrics											
Domestic	990.18	962.45	1 044.03	1 231.23	1 295.00	1 329.84	1 726.83	2 287.58	2 437.51	2 797.21	12.23
Import	na	na	na	na	na	na	15.77	12.55	49.31	10.55	
Export	51.00	77.39	95.54	88.40	84.35	89.63	89.68	90.99	157.64	197.24	16.22
Production	1 041.18	1 039.84	1 139.57	1 319.63	1 379.35	1 419.47	1 816.51	2 378.57	2 595.15	2 994.45	12.45
% Net export/ production	4.90	7.44	8.38	6.70	6.12	6.31	4.07	3.30	4.17	6.23	
Woollen fabrics											
Domestic	315.59	324.20	412.60	506.60	785.40	1 078.44	1 565.30	2 102.40	2 612.93	2 595.09	26.38
Import	na	na	na	na	na	na	13.05	31.70	82.97	9.10	
Export	9.79	19.36	23.95	22.03	21.20	24.20	28.64	41.28	51.49	70.00	24.43
Production	325.38	343.56	436.55	528.63	806.60	1 102.64	1 593.94	2 143.68	2 664.42	2 665.09	26.30
% Net export/ production	3.01	5.64	5.49	4.17	2.63	2.19	0.98	0.45	-1.18	2.29	
Knitting wool yarn											
Domestic	830.39	975.48	1 316.47	1 752.14	2 121.04	2 315.64	2 522.41	2 892.87	3 421.29	4 572.87	20.87
Import	na	na	na	na	na	na	44.53	104.22	147.70	22.43	
Export	0.00	0.00	0.86	4.43	3.78	11.35	8.00	32.66	66.85	237.04	123.00
Production	830.39	975.48	1 317.33	1 756.57	2 124.82	2 326.99	2 530.41	2 925.53	3 488.14	4 809.91	21.60
% Net export/ production	0.00	0.00	0.07	0.25	0.18	0.49	-1.44	-2.45	-2.32		

Table 6.3 (continued)

Sector	1978	1979	1980	1981	1982	1983	1984	1985	1986	1987	1978–87 average annual growth %
Woollen blankets											
Domestic	179.55	206.54	276.00	343.20	499.50	641.16	815.87	1 005.40	1 146.75	1 473.36	26.35
Import	na	na	na	na	na	na	3.26	0.53	1.29	0.14	
Export	37.26	39.75	50.95	58.04	62.39	70.00	63.44	55.22	120.86	113.03	13.12
Production	216.81	246.29	326.95	401.24	561.89	711.16	879.31	1 060.62	1 267.61	1 586.39	24.75
% Net export/ production	17.19	16.14	15.58	14.47	11.10	9.84	6.84	5.16	9.43	7.12	
Woollen knitted goods											
Domestic	352.95	483.75	590.90	575.36	555.15	507.03	594.34	761.30	859.20	930.56	11.37
Import	na	na	na	na	na	na	0.20	0.32	1.68	0.60	
Export	109.41	149.94	188.36	210.68	266.82	285.03	442.05	549.81	841.59	1 013.36	28.06
Production	462.36	633.69	779.26	786.04	821.97	792.06	1 036.39	1 311.11	1 700.79	1 943.92	17.30
% Net export/ production	23.66	23.66	24.17	26.80	32.46	35.99	42.63	41.91	49.38	52.10	
Average % net export / production	12.19	13.22	10.74	10.48	10.50	10.97	10.62	9.67	11.90	14.44	

Notes: na — not available.

a Domestic sale value is calculated according to the average industrial sale price. Export value is calculated by converting US$ price to RMB at the official exchange rate. Production is the sum of domestic sales and exports, that is, total output.

Sources: *China's Foreign Economic Relations and Trade Yearbook*, 1984: 915, 962, 963, 967–8, 970; 1985: 829, 834–6; 1968: 999, 1000, 1004–5, 1007; 1987: 351, 354–6. *China Statistical Yearbook*, 1981: 335, 391; 1984: 355–6, 390, 402; 1985: 512; 1987: 281, 284, 564, 602. *Chinese Textile Newspaper*, 13 June 1988, 16 June 1988.

woollen knitted goods, and knitting wool yarn. That is, the woollen fabrics subsector is the most labour-intensive and the knitting wool yarn sector is the most capital-intensive. The labour intensity of the blanket and worsted fabrics sectors are about average for the whole textile sector. Woollen fabrics have a labour intensity similar to cotton. Knitting wool yarn and knitted goods are relatively capital-intensive but none of the subsectors are as capital-intensive as the chemical fibre sector.

Based on these factor intensity data and given the characteristics of China's factor endowment noted above, we would have expected that the ranking of competitiveness of the subsectors would match that of their labour intensity; in other words, that the woollen fabrics subsector would be the most competitive and the knitting wool yarn subsector the least competitive, according to this criterion.

Subsector performance

Indicators of the performance of the industry, however, tell a different story. Output growth rates, export growth rates and ratios of export to output are summarised in Table 6.5. The subsectors are listed in that table in order of decreasing labour intensity. Also, as shown in Figure 6.3, the industry can be divided into two groups in terms of growth rates of output. The worsted subsector and the knitted goods subsector grew relatively slowly compared to the woollen

Figure 6.2 Sale value of the output of the wool textile sector, China, 1978–1987 (million yuan)

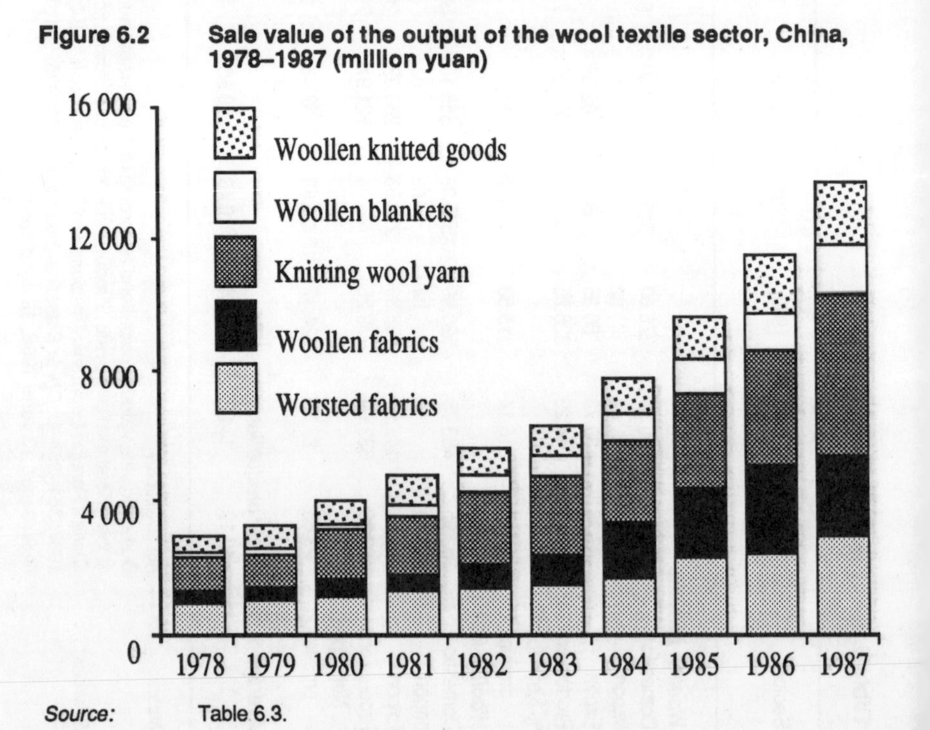

Source: Table 6.3.

Table 6.4 Wool textile industry indicators[a], China, 1978-87

Sector	1978	1979	1980	1981	1982	1983	1984	1985	1986	1987
Worsted fabrics										
Labour cost/value added %	10.00	9.80	9.90	8.90	10.00	11.60	11.70	16.70	21.00	21.70
Value added/sales value %	40.00	40.20	40.40	43.50	44.30	44.40	45.90	37.20	38.10	37.00
Value added/worker[b]	7 900	8 253	9 388	10 096	9 154	8 049	9 143	7 782	7 228	7 195
Profit+tax/fixed asset val. %	77.00	83.00	94.00	98.50	79.00	68.00	60.00	53.50	44.70	41.00
Profit+tax/output value %	32.40	35.00	35.30	36.80	35.00	34.00	34.00	31.70	29.30	26.20
Profit+tax/sales value %	33.00	34.00	34.40	35.90	35.00	34.15	33.30	28.30	26.00	23.40
Woollen fabrics										
Labour cost/value added %	16.00	16.20	18.20	19.40	19.00	23.00	25.94	29.90	29.80	32.00
Value added/sales value %	35.40	38.00	38.00	39.40	33.20	25.90	25.00	26.30	26.00	25.42
Value added/worker[b]	4 929	5 012	5 124	4 624	4 782	4 049	4 144	4 329	5 089	4 882
Profit+tax/fixed asset val. %	58.00	61.20	62.20	54.50	59.18	57.90	41.04	36.00	38.10	33.73
Profit+tax/output value %	31.60	34.00	34.00	34.00	32.50	30.00	25.00	22.50	19.50	17.62
Profit+tax/sales value %	31.60	32.50	32.50	32.50	31.00	27.90	22.73	20.00	17.80	16.09
Knitting wool yarn										
Labour cost/value added %	5.00	5.30	7.30	6.70	8.00	8.10	9.22	10.43	13.60	13.80
Value added/sales value %	36.40	34.00	34.90	35.90	32.20	29.80	29.50	29.10	29.20	26.30
Value added/worker[b]	15 547	15 247	12 759	13 505	11 357	11 502	11 660	11 229	11 100	11 296
Profit+tax/fixed asset val. %	158.00	160.00	133.00	137.00	103.00	105.00	86.25	80.36	66.80	62.80
Profit+tax/output value %	36.30	36.30	36.40	36.40	32.00	30.16	29.54	27.30	23.83	22.20
Profit+tax/sales value %	30.00	28.90	29.60	29.60	25.50	23.70	23.20	22.20	20.00	16.70
Woollen blankets										
Labour cost/value added %	11.00	11.20	12.20	14.20	14.60	15.62	15.67	17.00	19.30	22.08
Value added/sales value %	37.70	38.30	36.80	37.80	30.20	30.00	30.00	30.50	31.30	28.20
Value added/worker[b]	7 125	7 329	7 639	6 305	6 218	6 847	6 858	7 023	7 231	7 075
Profit+tax/fixed asset val. %	74.60	79.00	82.00	66.00	58.00	62.00	48.00	48.00	48.00	43.50
Profit+tax/output value %	31.00	33.00	33.00	26.70	26.70	26.70	26.70	26.70	26.70	22.00
Profit+tax/sales value %	31.00	32.60	31.30	31.30	23.80	23.00	22.00	21.42	21.40	18.00
Woollen knitted goods										
Labour cost/value added %	na	na	na	15.80	17.00	19.80	19.30	17.40	20.50	17.00
Value added/sales value %	na	na	na	32.80	37.90	36.40	39.70	44.60	45.70	50.00
Value added/worker[b]	na	na	na	5 688	5 365	4 712	4 919	4 371	7 046	10 245
Profit+tax/fixed asset val. %	na	na	na	141.00	89.00	66.00	62.00	72.50	59.80	43.60
Profit+tax/output value %	na	na	18.60	20.00	19.70	16.20	15.30	18.20	17.50	15.60
Profit+tax/sales value %	na	na	na	26.00	23.00	21.40	21.00	22.000	20.00	18.00
Wool textile industry										
Labour cost/value added %	9.90	10.68	11.88	11.69	12.47	16.21	16.20	16.29	18.69	21.22
Value added/sales value %	39.03	37.69	37.58	40.85	39.77	36.49	35.87	36.35	36.29	33.38
Value added/worker[b]	7 261	7 040	7 844	7 686	7 307	6 474	5 963	7 960	8 113	6 710
Profit+tax/fixed asset val. %	91.00	91.00	96.70	96.40	78.40	57.30	41.30	57.50	46.80	38.80
Profit+tax/output value %	31.70	33.64	34.35	33.78	30.44	23.88	24.40	28.25	26.15	23.67
Profit+tax/sales value %	32.18	32.06	31.94	33.74	31.39	26.24	26.02	27.64	24.78	21.13

Notes: na — not available.
a This table includes only enterprises under the Ministry of Textile Industry.
b Value added/worker unit is yuan/person.

Source: As for Table 6.2.

fabric and blanket subsectors. This is consistent with the ranking according to factor intensity. The outlier, however, is the knitting wool yarn sector, which, despite its capital intensity, grew relatively rapidly.

The other indicator shown in Table 6.5 is the degree of export orientation, measured by the ratio of exports to production. In all cases but one in the wool textile industry this ratio appears small (7 per cent at most in 1987). The exception is the relatively capital-intensive woollen knitted goods sector (52 per cent).

By comparison, Table 6.6 shows exports to production ratios for other parts of the textile industry. These data are defined in volume terms since comparable price data are not available. The export orientation in these sectors has risen to 7 per cent for cotton yarn and 19 per cent for cotton fabrics. The chemical fibre sector is a new importer. These data place the results for the wool textile sector in a clearer perspective. The wool textile sector overall is about as export orientated as the cotton yarn sector, more so than chemical fibres and less so compared to cotton fabrics. Its position on average by this criterion therefore is consistent with expectations based on factor intensity information.

In summary, while the labour-intensive woollen blanket and woollen fabric sectors have grown rapidly, so has the knitting wool yarn sector, which is the most capital-intensive of the group. Overall, the export orientation of the wool textile sector relative to other elements of the textile industry is consistent with expectations, except that the most export-oriented sector is the more capital-intensive woollen knitted goods sector. Another exception is that the sector with

Figure 6.3 Output growth index in the wool textile sector, China, 1978–87

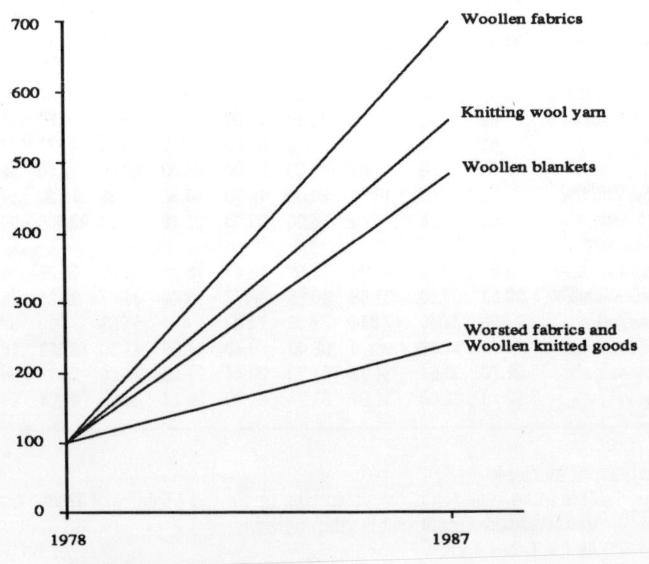

Note: Average annual output growth.
Source: Table 6.5.

Table 6.5 Wool textile industry, output and export growth rates, China, 1978-87 (per cent)

	Industry output value share 1987	Average annual output growth 1978-87	Average annual export growth 1978-87	Exports to production ratio 1978	1987
Woollen fabric	15.0	24.0	24.0	3.0	3.0
Woollen blankets	8.0	19.0	13.0	17.0	7.0
Worsted fabric	16.0	10.0	16.0	5.0	7.0
Wool knitted goods	7.0	10.0	28.0	24.0	52.0
Knitting wool yarn	28.0	21.0	123.0	0.0	5.0

Source: Author's calculations.

Table 6.6 Cotton yarn and fabric and chemical fibre production and exports, China, 1980-87

Sector	1980	1981	1982	1983	1984	1985	1986	1987
Cotton yarn (1000mt)								
Domestic	2 880.00	3 118.10	3 290.40	3 117.40	3 055.00	3 375.40	3 739.40	4 050.10
Imports			6.90	1.40	0.03	0.20	0.24	0.50
Exports	46.00	51.90	63.60	152.60	164.00	160.00	238.60	318.00
Production	2 926.00	3 170.00	3 354.30	3 270.00	3 219.00	3 535.00	3 987.00	4 368.10
% net export/ production	1.57	1.64	1.70	4.60	5.09	4.50	6.00	7.28
Cotton fabrics (million metres)								
Domestic	11 974.51	12 532.50	13 609.42	12 815.84	11 409.21	12 265.76	13 619.47	14 512.30
Imports			191.45	147.00	225.72	446.25	398.40	400.00
Exports	1 495.49	1 732.50	1 740.58	3 064.16	2 290.79	2 404.74	3 850.53	3 387.70
Production	13 470.00	14 270.00	15 350.00	15 880.00	13 700.70	14 670.50	17 470.00	17 900.00
% net export/ production	11.10	12.20	10.09	18.37	15.07	13.35	19.76	16.70
Chemical fibre (1000mt)								
Domestic	450.30	527.30	517.00	540.70	734.90	942.80	998.70	1147.90
Imports	411.40	629.00	434.40	397.70	520.50	831.40	475.20	370.00
Exports							18.68	27.10
Production	450.30	527.30	517.00	540.70	734.90	924.80	1 017.30	1 175.00
% net export/ production							-44.90	-29.00

Source: *China Statistical Yearbook*, 1988.

the fastest export growth, relative to output growth, is the knitting wool yarn sector, which is also relatively capital-intensive. These results suggest there are factors influencing performance other than the intensity of labour use.

The next section explores the influence of some special features of processing industries, in particular, constraints on the supply of raw materials.

RAW MATERIAL SUPPLY ISSUES AND RESPONSES

Raw wool import penetration

The Chinese wool textile sector consumes both imported raw wool and raw wool produced in China. As noted in Chapter 3, China is a net importer of raw wool. The availability of foreign exchange to buy imports is constrained by the overvaluation of the exchange rate, which reduces the supply generated by exports. There were two mechanisms by which that foreign exchange could be allocated in the period since 1978.

Before the emergence of the secondary markets in 1988, a particular sector or individual enterprise would obtain foreign exchange either from an allocation from the planning system or else would draw upon their funds retained from export sales. After the emergence of the secondary markets, this quantity constraint on access to foreign exchange was replaced by a market process. Enterprises could bid for foreign exchange to buy imports, but that bidding process added a premium to the price of foreign exchange, as explained in Chapter 3.

In summary, wool imports can be thought of as either being restricted by quantity constraints imposed by the planning system or, more recently, as a result of importers being faced with the premium on foreign exchange evident in the secondary market.

Table 6.7 illustrates the importance of raw wool imports in China's total raw wool consumption in the period up to 1987. The rapid growth of textile industry output has seen the penetration of imported wool rise from about 15 per cent of consumption in 1978 to 62 per cent in 1987. The rate of increase of import penetration is illustrated in Figure 6.4. This occurred despite the control of imports by the planning system.

The data in Table 6.7 are reported in volume terms but some allowance for differences in quality between imported and domestic wool has been made by calculating import penetration in terms of scoured wool. The degree of substitutability between wool of different origins depends on the type of product. Some sectors require high quality (low fibre diameter) wool which is available only in the import market. The importance of this constraint can be illustrated by reviewing the experience of subsectors of the wool textile industry, beginning with the type of wool consumed by each.

Quality composition of wool consumption and sectoral growth rates

The quality composition of raw wool consumption is shown in Tables 6.8 and 6.9 and illustrated in Figure 6.5. These data can be compared to quality mix of

Table 6.7 Scoured wool consumption, China, 1978-89 (tonnes)

	Origin		
	Domestic	Import	Total[b]
1978	65 909	11 529 (15)[a]	77 438
1979	66 373	12 683 (16)	79 056
1980	72 082	20 901 (23)	92 983
1981	70 562	37 744 (35)	108 306
1982	81 410	55 790 (41)	137 200
1983	74 710	74 060 (50)	148 770
1984	77 000	75 000 (49)	152 000
1985	72 000	91 000 (56)	163 000
1986	66 000	110 000 (63)	176 000
1987	70 000	113 000 (62)	185 000
1988	74 253	138 700 (65)	212 953
1989	79 713	77 250	na

Notes: na — not available.
a Numbers in parentheses represent ratio of imports to total.
b No allowance is made for changes in stocks.

Sources: IWS, *Wool Facts*, 1984, 1987 1988: *China Textile Statistics; China Statistical Yearbook*, 1989: *China Customs Statistics*, 1989-1, 1990-1.

Figure 6.4 Scoured wool consumption, China, 1978–87 (tonnes)

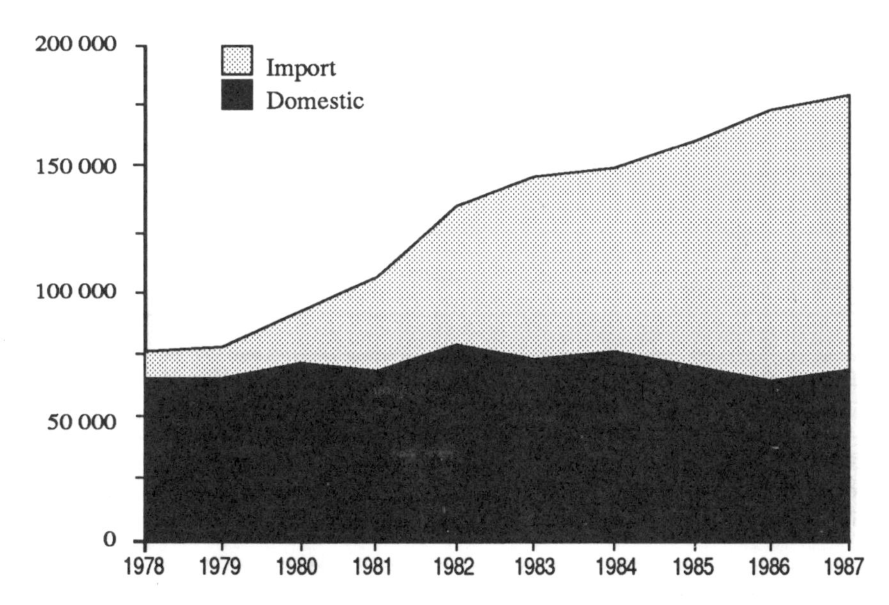

Sources: IWS, *Wool Facts*, 1984–1988; Ministry of Textile Industry, *Textile Industry Information*, 1978–1988.

Chinese raw wool production in Table 6.10, summarised in Figure 6.6.

Fine wools are used in worsted fabrics and some high quality woollen fabrics, as well as high grade woollen knitted goods, and some knitting yarn, mainly for machine knitting. Semi-fine wool is consumed in most sectors. Coarse wool is used in blankets, carpets, woollen fabrics, hand-knitting yarns and some woollen knitted goods.

Even though the share of fine wool in total raw wool production is significant, Chinese fine wool suffers from other problems, such as breaks in the fibre and high levels of dirt and vegetable matter, as indicated by the low scoured yields (see Chapter 8 for more details). Of the three product types which use relatively more fine wool (worsted, woollen fabrics and knitted goods), the woollen subsector is relatively less sensitive to these other impurities. Given the foreign exchange constraint, the supply of raw wool of appropriate quality will therefore have constrained the growth of the other subsectors.

Some elaboration of this argument is required for the knitted goods sector. That sector (as defined in the sources of the statistics) produces no yarn of its own but consumes the output of the knitting wool yarn sector or else imports yarn.

The demand for high quality wool by the knitting wool yarn sector has increased since 1978. Table 6.11 shows the composition of output of the knitting yarn sector. In 1978 about 75 per cent of the industry's output was yarn for hand knitting, which requires coarser wool (see Table 6.11 footnote for details). By 1987 this share had fallen to 49 per cent and output of yarn for machine knitting, which requires much finer wool, had grown by a factor of nearly 11. The export

Table 6.8 Quality composition of raw wool consumption by sector, 1986 (per cent)

Sector wool quality	70s (18.1-20.1u)	66s (20.1-21.5u)	64s (21.6-23u)	60s (23.1-25u)	58s (25.1-27u)
Worsted fabric	10	30	50	10	
Woollen fabric	5	10	30	35	10
Knitting wool yarn				5	10
Woollen blankets					
Woollen knitted goods	3	5	20	40	15

	56s (27.1-29.5u)	50s (29.6-32.5u)	48s (32.6-35.5u)	46s (35.6-38.5u)	44s (38.6-42u)
Worsted fabric					
Woollen fabric	10				
Knitting wool yarn	15	30	30	10	
Woollen blankets			25	35	40
Woollen knitted goods	10	7			

Source: Author's estimates.

Figure 6.5 Wool consumption by quality number, China, 1986, 1987 (per cent)

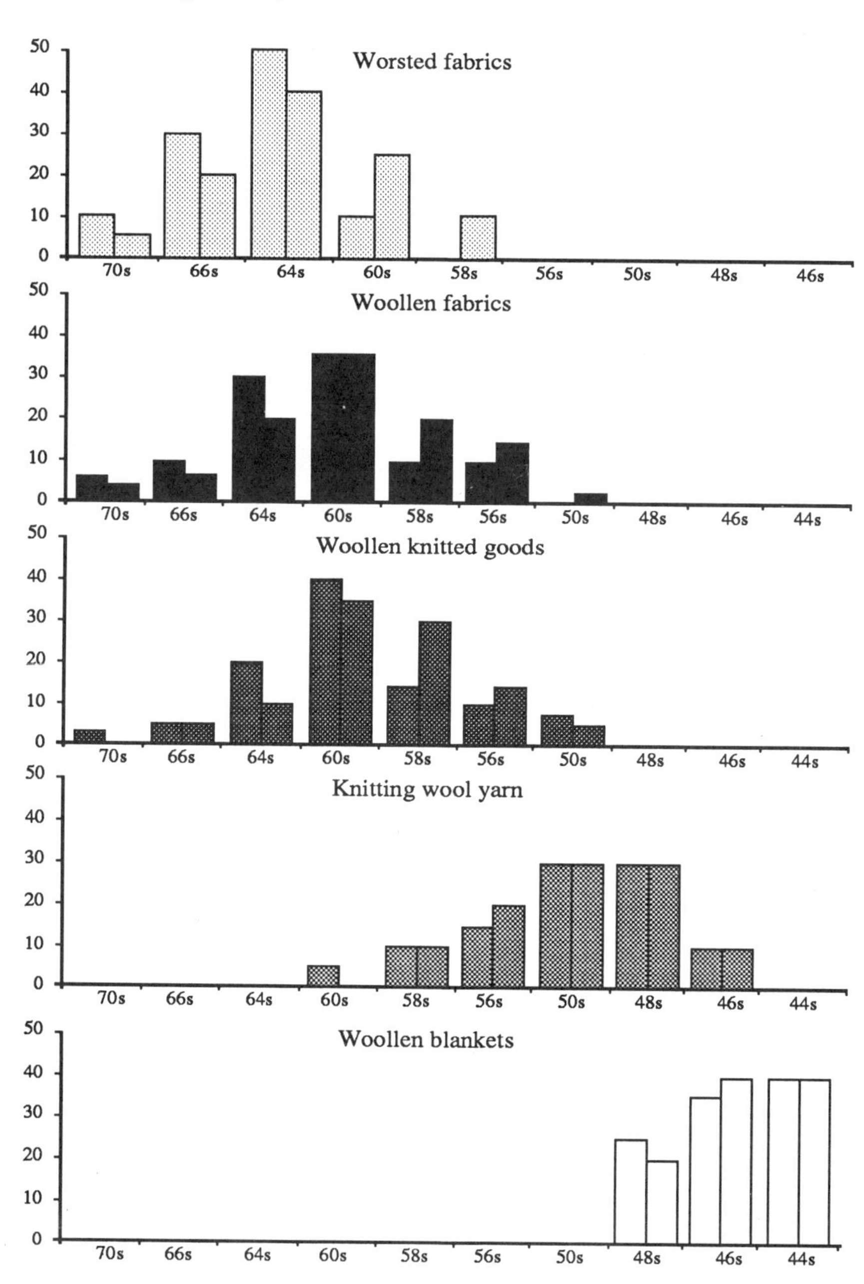

Sources: Tables 6.8, 6.9.

Figure 6.6 Chinese wool production, 1980–87 (tonnes, greasy basis)

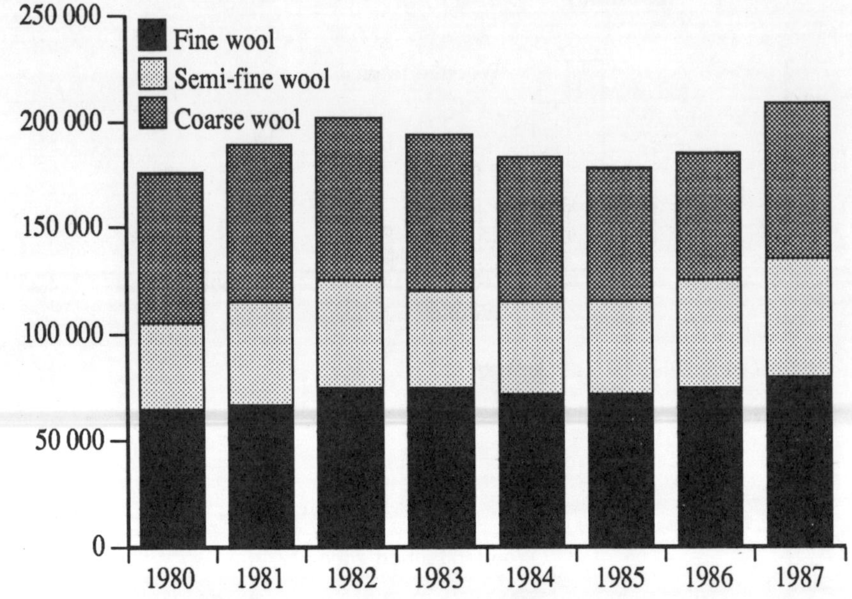

Sources: *Chinese Agricultural Yearbook*, 1980–88; *China Statistical Yearbook*, 1988.

Table 6.9 Quality number of raw wool consumption by sector, 1987 (per cent)

Sector wool quality	70s (18.1-20.1u)	66s (20.1-21.5u)	64s (21.6-23u)	60s (23.1-25u)	58s (25.1-27u)
Worsted fabric	5	20	40	25	10
Woollen fabric	3	5	20	35	20
Knitting wool yarn				5	10
Woollen blankets					
Woollen knitted goods		5	10	35	30

	56s (27.1-29.5u)	50s (29.6-32.5u)	48s (32.6-35.5u)	46s (35.6-38.5u)	44s (38.6-42u)
Worsted fabric					
Woollen fabric	15	2			
Knitting wool yarn	20	30	30	10	
Woollen blankets			20	40	40
Woollen knitted goods	15	5			

Source: Author's estimates.

Table 6.10 Quality composition of raw wool production, China, 1980–87 (tonnes)

Range	1978	1979	1980	1981	1982	1983	1984	1985	1986	1987
Total raw wool										
Quantity	146 000	157 000	157 000	189 000	202 000	194 000	183 000	177 953	185 196	208 909
%	100	100	100	100	100	100	100	100	100	100
Fine wool (60s-70s)										
Quantity			69 000	74 722	88 388	88 949	85 548	85 860	89 758	100 059
%			39.3	39.5	43.8	45.8	46.7	48.2	48.5	47.9
Semi-fine wool (46s-58s)										
Quantity			36 903	39 498	41 720	38 008	34 384	32 070	33 335	37 000
%			21.0	20.9	20.7	19.6	18.8	18	18	17.7
Sub-total (%)			60.3	60.4	64.5	65.4	65.5	66.2	66.5	65.6
Coarse wool (below 46s)										
Quantity			69 825	74 843	71 730	67 136	63 069	60 022	62 103	71 850
%			39.7	39.6	35.5	35.5	34.5	33.8	33.5	34.4

Note: 60s-70s: 25u-19.1u; 46s-58s: 38.5u-25.1u; below 46s: 38.6u-42u.

Sources: *Almanac of China's Agriculture*, 1980: 120, 124; 1984: 117; 1986: 98, 213. *Almanac of China's Rural Statistics*, 1986: 98, 108. *Almanac of China's Statistics*, 1987: 179, 183; 1988: 265.

of knitted goods has grown rapidly since 1978 and, as noted in Table 6.5, the knitted goods sector is highly export-oriented. This relieves the foreign exchange constraint for that sector, either to buy and import its own yarn or else to provide the foreign exchange to assist the yarn producers to buy the imported wool required. Thus the export orientation of the knitted goods sector has indirectly facilitated the growth of the yarn sector, especially that part which produces yarn for machine knitting, despite the relatively high degree of capital intensity of yarn production.

The argument of this section has been that the constraints on raw wool supply have had a differential impact on different sectors of the industry and so the composition of its output has changed. Another form of adjustment is evident in Figure 6.5 and that is the substitution into the use of coarser wools. This is evident especially in the fabric sectors and the knitted goods sector where finer wools are used.

Substitution into chemical fibres

Another response to the supply constraints facing the industry has been to substitute between types of fibres. Substitution has taken two forms, one the greater use of chemical fibres and the other the shift into the use of coarser wool.

Table 6.12 (and Figure 6.7) shows the use of chemical fibres in the wool textile industry. In 1978 chemical fibres accounted for about 35 per cent of total fibre consumption. This rose to nearly 50 per cent by 1987. In terms of volume, this substitution amounted to just over 50 mkg of raw material.

Table 6.11 Composition of output of the knitting yarn sector[a], China, 1978-87 (1000 metres)

Sector	1978	1979	1980	1981	1982	1983	1984	1985	1986	1987
Hand coarse knitting wool yarn	26.30	29.30	33.10	38.90	49.50	54.50	55.60	54.80	62.00	84.70
Hand fine knitting wool yarn	1.00	2.80	3.80	6.20	6.80	6.60	8.80	11.30	10.20	14.30
Knitting wool yarn for knitted goods	4.60	12.40	20.40	32.00	36.20	40.90	45.50	58.40	76.40	100.40

Note: a Hand coarse knitting wool yarn uses 48s-58s quality number wool or 5-10D. chemical fibre. Hand fine knitting wool yarn uses 60s-64s quality number wool or 5-10D. chemical fibre. Knitting wool yarn for knitted goods uses 64s quality number wool or 3-6D. chemical fibre.

Source: As for Table 6.2.

Decline in rate of capacity utilisation

Table 6.13 shows the level of capacity utilisation by subsector, based on estimated spindle requirements to produce the observed yarn output. The overall level of capacity utilisation (spindles only) fell from 95 per cent in 1978 to 77 per cent in 1987. The greatest decline was in the worsted sector, from 91 per cent to 65 per cent. This result is consistent with the earlier argument that this sector was the one most constrained by the raw material supply problems.

Table 6.12 Chemical fibre consumption in the wool textile industry, China, 1978-87 (tonnes)

Sector	1978	1979	1980	1981	1982	1983	1984	1985	1986	1987
Total raw material consumption										
Quantity	119 873	131 322	155 783	192 032	233 610	253 722	254 900	286 200	337 800	362 380
%	100	100	100	100	100	100	100	100	100	100
Wool consumption										
Quantity	77 438	79 056	92 983	108 306	137 110	148 722	152 000	163 000	176 000	183 000
%	64.60	60.20	59.70	56.40	58.70	58.60	59.60	57.00	52.00	50.50
Chemical fibre consumption										
Quantity	42 435	52 266	62 800	83 726	96 500	105 000	102 900	123 200	161 800	179 380
%	35.40	39.80	40.30	43.60	41.30	41.40	40.40	43.00	48.00	49.50

Source: *China's Textile Industry Yearbook*, 1983: 290; 1984–85: 398; 1986–87: 378.

Figure 6.7 Chemical fibre vs. scoured wool consumption in the wool textile sector, China, 1978–87 (tonnes)

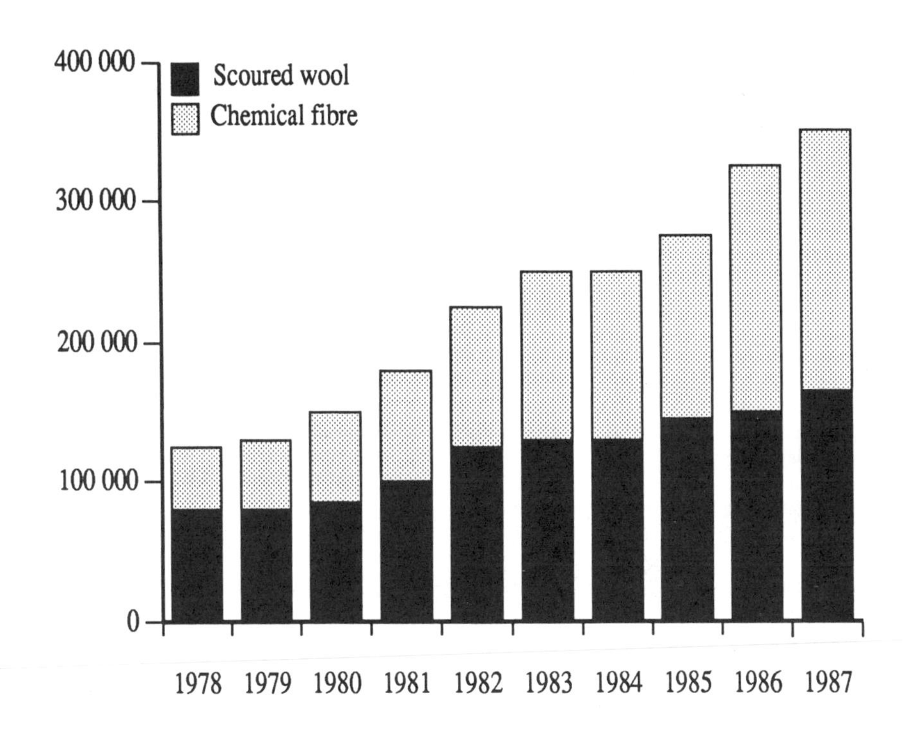

Sources: *China Textile Industry Yearbook*, 1983–87; and calculated.

The excess capacity in the industry is significant in absolute terms. For example, had it been possible to maintain utilisation rates at 95 per cent, total scoured wool consumption would have been about 10 per cent higher than the actual consumption level in 1987. At a yield of 55 per cent, this difference is equivalent to just over 30 mkg of greasy wool.

Growth in township capacity

The decline in capacity utilisation also reflects the investment strategies of local governments. From Figure 6.3 it is evident that the fastest growing sector of industry was woollen fabrics and the slowest growing (in output terms) was worsted fabrics. Central government policy, since 1982, released many regions from the control of the development plan of the Ministry of Textile Industry. As a result, and especially in Jiangsu, Shandong and Zhejiang provinces, the woollen textile industry grew rapidly.

Table 6.13 Capacity utilisation in the wool textile industry, China, 1978-87

Sector	1978	1979	1980	1981	1982	1983	1984	1985	1986	1987
Worsted fabrics										
Yarn prod (m)	26 900	26 680	29 060	32 140	32 050	32 830	39 270	45 580	49 360	54 970
Spindle requirements[a]	26.24	26.15	28.35	31.36	31.27	32.08	38.31	44.47	48.16	53.63
% spindle utilised	91.00	89.00	96.00	95.60	83.00	72.00	77.00	73.30	66.00	65.30
Woollen fabrics										
Yarn prod (m)	18 820	20 330	24 350	28 110	39 250	50 860	68 620	86 470	10 6490	10 4790
Spindle requirements[a]	5.81	6.27	7.52	8.68	12.12	15.70	21.18	26.69	32.87	32.54
% spindle utilised	77.00	75.00	72.00	72.00	73.00	75.00	75.00	70.30	83.00	80.00
Knitting wool yarn										
Yarn prod (m)	37 800	44 400	57 300	76 500	92 500	102 100	110 000	125 900	149 100	204 232
Spindle requirements[a]	10.80	12.69	16.37	21.86	26.43	29.17	31.43	35.97	42.60	58.35
% spindle utilised	100.00	100.00	93.00	95.60	88.00	94.60	96.70	97.60	81.00	91.30
Woollen blankets										
Yarn prod (m)	11 563	12 765	16 348	19 740	22 508	29 999	32 290	37 278	44 807	55 833
Spindle requirements[a]	1.78	1.96	2.52	3.04	4.61	4.90	5.12	5.34	6.89	8.59
% spindle utilised	93.70	92.50	87.60	77.50	80.30	97.80	95.40	85.30	92.00	96.50
Woollen knitted goods										
Yarn prod (m)	na	na	na	6 062	5 802	5 485	6 913	9 294	10 352	9 832
Spindle requirements[a]	na	na	na	2.80	2.70	2.53	3.19	4.29	4.78	4.54
% spindle utilised	na	na	na	97.00	85.00	75.00	88.60	93.00	89.20	79.00
Total spindle req.	44.71	47.07	54.76	64.94	72.52	81.80	96.04	112.47	130.52	152.91
Total spindle capacity	47.81	53.30	60.05	74.44	88.90	100.53	120.52	139.49	168.53	199.20
% spindle utilisation	93.50	88.30	91.20	87.20	82.00	81.00	80.00	81.00	77.40	76.70

Notes: na — not available.
a Spindle requirements unit: 10,000 spindles.
Source: *China Textile Industry Yearbook*, 1984-85, 1986-87; *China Textile Newspaper*, 11 July 1988.

Table 6.14 illustrates the increasing importance of East China as the location of the industry. By 1987, the three provinces, and Shanghai, in that region accounted for nearly half the spindles in China, and most of the growth was due to expansion in Jiangsu. That province alone accounts for nearly a quarter of the total (see Figure 6.8). Much of the capacity increase was also organised outside the state system by township textile enterprises. Table 6.15 shows the growth in wool textile production in township enterprises since 1985. By 1987 this sector accounted for about 38 per cent of the industry total, up from 19 per cent two years before.

One reason for the growth of woollen fabric production in the township sector was the level of demand. In 1978 the ratio of the number of woollen spindles to worsted spindles was about 1:4. As a result, woollen products were relatively scarce compared to worsted ones, and there was an incentive to develop the capacity of the woollen fabric sector.

In addition, as noted above, the woollen sector is relatively more labour-intensive and thus for township enterprises, and even urban enterprises, the

Figure 6.8 Wool textile spindle numbers by province, China, 1983, 1987

Total number: 1.01 million

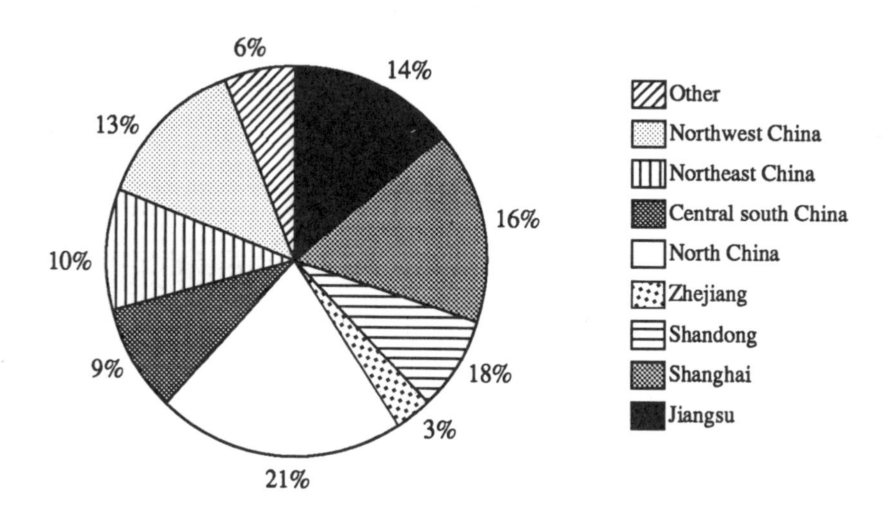

Total number: 1.99 million

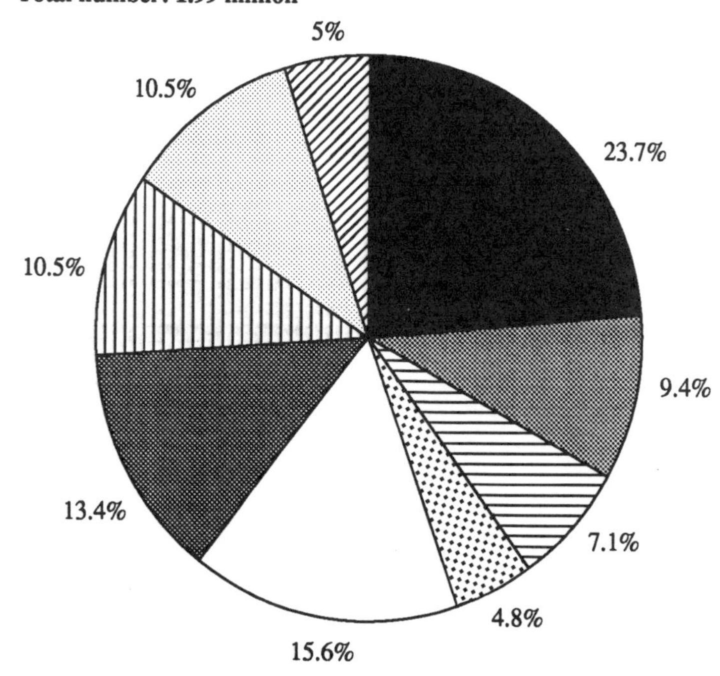

Source: *China Textile Newspaper*, 14 July 1988.

Table 6.14 Wool textile industry by region, China, 1983, 1987 (1000 spindles)

Region	1983	%	1987	%
China total	1 005.30	100.00	1992.00	100
Worsted spindle	445.10		821.00	
Woollen spindle	256.70		489.00	
Knitting yarn spindle	308.50		639.00	
East China	433.30	43.00	975.90	
Worsted spindle			360.50	
Woollen spindle			210.80	
Knitting yarn spindle			295.90	
Jiangsu	139.90	14.00	472.10	23.70
Worsted spindle			159.80	
Woollen spindle			143.00	
Knitting yarn spindle			157.00	
Shandong	78.90	8.00	142.20	7.10
Worsted spindle			53.40	
Woollen spindle			33.60	
Knitting yarn spindle			48.50	
Shanghai	163.30	16.00	187.80	9.40
Worsted spindle			105.90	
Woollen spindle			23.50	
Knitting yarn spindle			58.40	
Zhejian	32.00	3.00	95.30	4.80
Worsted spindle			22.60	
Woollen spindle			21.70	
Knitting yarn spindle			44.50	
North China	206.40	21.00	310.00	15.60
Worsted spindle			151.90	
Woollen spindle			63.90	
Knitting yarn spindle			33.50	
Beijing	67.20	7.00	88.10	4.40
Worsted spindle			43.50	
Woollen spindle			9.60	
Knitting yarn spindle			33.50	
Inner Mongolia	36.50	4.00	58.00	2.90
Worsted spindle			28.10	
Woollen spindle			14.20	
Knitting yarn spindle			15.80	
Tianjin	63.60	6.00	93.40	4.70
Worsted spindle			53.90	
Woollen spindle			17.90	
Knitting yarn spindle			19.80	
Central Sth China	86.50		266.50	
Worsted spindle			79.40	
Woollen spindle			73.60	
Knitting yarn spindle			78.00	
Henan	37.50	4.00	81.20	4.10
Worsted spindle			25.80	
Woollen spindle			36.60	
Knitting yarn spindle			17.30	
Hubei	35.90	4.00	85.70	4.30
Worsted spindle			32.60	
Woollen spindle			16.90	
Knitting yarn spindle			32.40	
Northeast China	104.00	10.00	208.80	10.50
Worsted spindle			96.10	
Woollen spindle			50.50	
Knitting yarn spindle			61.80	
Heilongjiang	22.30	2.00	66.40	3.30
Worsted spindle			26.00	
Woollen spindle			16.50	
Knitting yarn spindle			23.90	
Liaoning	55.50	6.00	100.90	5.10
Worsted spindle			49.40	
Woollen spindle			24.80	
Knitting yarn spindle			26.40	
Northwest China	131.00	13.00	208.50	10.50
Worsted spindle			116.20	
Woollen spindle			42.20	
Knitting yarn spindle			49.20	
Gansu	47.80	5.00	65.00	3.30
Worsted spindle			30.70	
Woollen spindle			17.00	
Knitting yarn spindle			17.30	
Xinjiang	33.10	3.00	61.70	3.10
Worsted spindle			39.30	
Woollen spindle			9.20	
Knitting yarn spindle			13.10	

Note: Some spindle types are not listed here.
Sources: *Almanac of China's Textile Industry*, 1984-1985: 409, 413; *China Textile Newspaper*, 14 January 1988, 11 July 1988, 14 July 1988.

Table 6.15 The township wool textile industry, China, 1985, 1987

	Number of enterprises		Staff & workers (1000 persons)		Output value (million yuan)	
	1985	1987	1985	1987	1985	1987
Textile	23 129	28 848	1 900	2 735.2	16 668	31 590
Cotton textile	5 792	7 162	622.2	800	5 851	11 678
Wool textile	1 452	1 988	169.8	283	1 804	6 191
Cotton knitted goods	7 158	8 317	545.1	635	3 078	5 275

Sources: *China Textile Newspaper*, 13 September 1986, 9 June 1988; *China Statistical Yearbook*, 1987: 265-6; 1988: 322-5.

development of woollen spinning was more profitable than worsted production. However, the passing down of autonomy to the local level, and the independent investment decisions, led to rapid growth in raw wool demand, not all of which could be supplied, for reasons discussed above.

While the country as a whole is a net wool importer, the position varies between regions. Comparing regional shares in spindle numbers with regional shares of greasy wool production, the big wool producing regions in North and Northwest China continue to be net exporters while East China is the major net importing region. The regulation of the raw wool market, and the rents existing in processing, create strong incentives for wool producing areas to withhold supplies of raw wool for local processing, as discussed in relation to Figure 6.4. This has exacerbated the raw material supply problem facing the rapidly growing industry in the eastern provinces, and led to the phenomenon of the 'wool war' (see Chapter 9).

A further effect of this process of adjustment to raw material supply problems has been a dramatic decline in profitability. Capacity utilisation has fallen, but, in the short run, labour has not been (or cannot be) laid off. Also, wages have been rising. The average wage paid to staff and workers in the textile industry rose from 707 yuan in 1980 to 1,250 yuan in 1987, a rise of 74 per cent. Wool prices, especially outside the state procurement system, have also been rising. As a result, the share of value added in sale value has risen more than twice; and the share of profit and tax in sale value has fallen from over 30 per cent to just over 20 per cent.

CONCLUSIONS

The wool textile sector has grown rapidly since the reforms of 1978. Its growth rate of output exceeded that of the rest of the textile sector. This performance reflects its competitiveness in terms of degree of labour intensity. However, a thorough understanding of the performance of the wool textile sector requires its disaggregation into the major components, such as woollen fabrics, woollen blankets, worsted fabrics, woollen knitted goods and knitting yarn. At this level of disaggregation, the performance of each subsector varies considerably. In part, this variation reflects differences in labour intensity. But, it was argued, more important has been the ability of the subsectors to cope with constraints in raw material supplies.

In a free-trade regime, raw material supplies need not be an issue. But in the Chinese system, when foreign exchange rationing acts like a quota in imports, it becomes one. Subsectors are hit to different degrees depending on the quality of wool required and its local availability. As a result, the worsted sector, for example, has been severely constrained by exploiting a high degree of export orientation in their production. This appears not to be the case for the knitted goods producers and their yarn suppliers.

In the medium term, given the type of wool they require, domestic raw wool supplies are likely to be dedicated to the woollen fabric sectors, production of hand-knitting yarn and blankets. This is already evident in the higher rates of output growth of these sectors and their relatively higher rates of capacity utilisation. However, the reliance of imports of fine wool, and the rapid rise in the level of penetration of imports into raw wool consumption in China, will contribute to the interest in raising the level of self-sufficiency in fine wool consumption. Pursuit of these strategies, however, raises a number of issues in relation to breeding programs in the sheep industry, which are reviewed in Chapter 8.

The replacement of an administrative procedure for allocating foreign exchange by a secondary market improves the efficiency of the allocation of foreign exchange. Import purchases at the margin are reduced because of their relatively high price. The impact of this surcharge between sectors still depends on the degree of substitutability between domestic and imported wool.

A couple of other issues were noted in this chapter. One is the rising cost of labour to the wool textile industry. As argued in Chapter 5, this will continue as long as China continues to industrialise. The pressures of higher labour costs are evident in the financial indications of the sector reported in this chapter, even since 1978. The chapter also highlighted the very rapid growth of the textile industry in the Chinese countryside. This development has added to the demand for raw wool and put new pressures on the raw wool marketing system. These pressures and the responses to them are explored in more detail in Chapter 9.

7 China's export marketing performance and the pressures for reform

JAMES CROWLEY, CHRISTOPHER FINDLAY AND
MELISSA GIBBS

INTRODUCTION

A common approach to the analysis of prospects for exports of textiles and clothing is to identify the intensity with which labour is used in the production process and then examine the national endowment of labour relative to the exporting country's competitors (see Anderson and Park, 1989, for example). The point has also been made more recently in relation to merchandise trade that market penetration may not only depend on competitiveness in the process of production of the basic item but could also depend on the competitiveness of the provision of complementary services (Jones and Kierzkowski, 1989). The relevance of the availability of these traded services is often illustrated by examples of investment goods or consumer durables, such as motor cars.

In this chapter, it is argued that while the factor intensity of the production process is a powerful determinant of competitiveness in the textile and clothing trade, the availability of complementary services is also critical in this industry. The chapter reports one of the first surveys to measure the extent of the importance of suppliers' competitiveness across a range of product and service attributes from the point of view of the importer.

Attention is focused on the performance of China as a textile and clothing (TC) exporter. The analysis is especially relevant to China because, as argued in Chapter 5, China will be highly competitive in TC exports. The extent to which China can match competitiveness in the production process with competitiveness in complementary services will influence the rate of growth of China's share of the world market. China's performance on these production and other service criteria is compared with that of members of the newly industrialising and other Asian developing country groups.

One problem in evaluating the performance of various countries in these dimensions is the interaction with the protection arrangements for textiles and clothing. World markets in these items are dominated by quota controls which are specific to particular supplying countries and so the influence of the various elements of competitiveness will not be fully reflected in market shares.[1] While these arrangements permit quota holders to export up to quota limits, thereby potentially promoting or maintaining market share of less competitive suppliers, they may also create a bias against newcomer suppliers (Hamilton, 1986). These issues are discussed in more detail in the introduction to this volume.

The Australian market is an exception to the system of country specific quotas. Australia has a global rather than a country specific quota system for TC imports. In recent years, TC imports have been restricted by policy to about 20 per cent of the Australian consumption of clothing and about a third of the consumption of textiles. Australian textile imports are generally restricted by tariffs but not quotas. In the case of clothing there is a two-tier tariff, the first level applying up to a certain quota level of imports and the second level applying to imports over the quota. This has the effect of limiting imports close to the quota level. The total import volume is first allocated between Australian importing firms (partly on a competitive basis through a tender system) and then, in a second stage, by those firms to exporting countries according to commercial decision-making. Shares in the Australian import market of countries of interest there are shown in Tables 7.1 and 7.2.

Table 7.1 Australian imports of textiles from Asia and the rest of the world, 1980, 1983, 1986, 1987 (per cent)

	1980	1983	1986	1987
NIEs				
Hong Kong	8.48	7.04	6.62	7.60
Korea	4.66	6.55	7.35	8.50
Taiwan	5.09	8.88	10.03	11.84
Singapore	1.34	1.03	0.46	0.44
Total	19.57	23.50	25.06	28.30
	[1.97][a]		[1.69]	
Other Asian developing				
China	6.72	6.64	6.99	7.81
India	4.18	2.15	1.50	1.48
Pakistan	0.88	1.60	2.24	2.66
Other ASEAN	2.99	3.46	5.34	6.57
Total	14.77	13.85	16.07	18.52
	[1.48]		[1.37]	
Rest of world				
	65.65	62.67	58.86	53.11
Total	100.00	100.00	100.00	100.00

Note: a Unadjusted intensity indexes in square brackets.
Source: International Economic Data Bank, Australian National University, Canberra.

Table 7.2 Australian imports of clothing from Asia and the rest of the world, 1980, 1983, 1986, 1987 (per cent)

	1980	1983	1986	1987
NIEs				
Hong Kong	22.99	21.73	20.61	18.19
Korea	4.84	9.72	9.91	10.77
Taiwan	18.82	20.78	14.43	14.49
Singapore	1.15	0.38	0.17	0.21
Total	47.80	52.61	45.12	43.66
	[1.78][a]		[1.57]	
Other Asian developing				
China	15.74	14.33	17.66	21.44
India	3.02	3.30	3.90	3.84
Pakistan	0.45	0.27	0.28	0.26
Other ASEAN	8.83	5.48	4.71	4.55
Total	28.04	23.38	26.55	30.09
	[3.10]		[1.74]	
Rest of world				
	24.17	24.02	28.34	26.24
Total	100.00	100.00	100.00	100.00

Note: a Unadjusted intensity indexes in square brackets.
Source: International Economic Data Bank, Australian National University, Canberra.

Textiles refers to cloth, and clothing to made-up garments. The first point evident in Tables 7.1 and 7.2 is the lower degree of penetration by the newly industrialising economies (NIEs) and other Asian developing economies in the textile market compared to the clothing market. One explanation for this is that textiles production tends to be more capital-intensive than the assembly of garments. As a result, the trend is for the group of countries of interest to be more competitive suppliers of clothing than of textiles. This result is qualified by the wide variety of types of products within each category.

In the Australian market for textile imports, the NIEs' share has risen rapidly, although not quite as fast as their share in world textile exports (as shown by the drop in the ratio of their share in Australian imports to their share in world exports, which is defined as 'the unadjusted intensity coefficient' shown in the tables). The share of other developing Asian economies in Australian imports has moved in parallel with their share of world trade.

In the clothing market the share of the developing countries in the Australian market has fallen, and the intensity coefficient nearly halved. The biggest suppliers in this group in the clothing market, as well as the textile market, were China and ASEAN (other than Singapore). The most substantial change in both tables has been China's penetration of the Australian clothing market.

The varying levels of success in penetrating the Australian market are of special interest here. Since Australia's protective arrangements are not country-specific, import success is a direct outcome of competitiveness. Thus an opportunity is presented to study the relative competitiveness of the different exporters, which is not presented in import markets for which country-specific quotas are in operation.

In the next section, some recent developments in the logistics chain for the production and export of textiles and clothing are reviewed and the relevant set of dimensions of the product, apart from cost at factory door, which influence competitiveness are defined. Results of the survey of Australian importers are presented in the third section. A fourth section contains discussion of some implications of the results.

LOGISTICS CHAIN

The idea that there are characteristics of a product, other than those it has at the factory door, which affect competitiveness is an old one in economics. Textbooks point out that value is defined in terms of time, place and form, for example. The issues of product attributes and the competitiveness of their creation has also received a great deal of attention from management researchers. The possible trade-offs between product attributes and the impact on profits of those choices has been the subject of 'logistics management' research.

The management of logistics involves a search for the optimal balance between the competing elements of production cost, inventory cost, transport cost and revenue losses associated with poor customer service. Firms over-stressing the importance of responding to customer demands tend to deliver a high quality of service by incurring high transport and retail stock holding costs. Firms concentrating on transport cost minimisation tend to allow this to dominate customer delivery times and production schedules. Organisations with a strong emphasis on financial variables tend to concentrate on inventory cost minimisation with resultant difficulties for the schedulers of transport and production. Organisations which are production dominated often overrule customer preference and transport considerations by criteria such as minimum production cost.

There are trade-offs between meeting the customer's specifications of the preferred product, transport costs, inventory costs and production costs. Over time, and as a result of economic development, the optimal choice of the mix of characteristics in a product for a particular market will alter. For example, a phenomenon often observed in the industrialised countries is the increasing degree to which consumers can specify the characteristics of goods they purchase, including durable goods like motor cars (Kumpe and Bolwijn, 1988). Developments of this type have significant implications for the management of the production process, the distribution system and the inventory systems. In other economies, at earlier stages of development, the stress may be placed on production and scale rather than these other aspects of distribution and accommodation of the individual preferences of consumers.

The field of 'logistics management' covers a very large number of decisions such as those on the location of production, the choice of transport mode, inventory size and size of production run. In addition, there are decisions about flows of information and finance which complement those concerning goods flows. These too are included in the scope of logistics.

In some cases, particular choices in each of these dimensions are considered under titles such as materials requirements planning or just-in-time inventory management systems. In the textile and clothing industry, the strategy package is called 'quick response' (QR).[2] This strategy has as its objective a much quicker response to customer orders for TC products of particular types, which is achieved by generally lower average lot sizes, reduced lead-times (between the date of placing an order and receiving the product), and leads to greater use of smaller unit size transport systems, for example.

Clearly, there is some overlap between the considerations of logistics and elements of competitiveness sometimes identified under other headings. But in this chapter, rather than talk in terms of 'marketing variables' or 'trade resistances' or the elements of the 'production chain', we develop our arguments in terms of the elements of logistics, which in our view have a more specific meaning and greater relevance for the purpose of this chapter.

The broad elements of competitiveness from the logistics perspective will already be evident from the definition of logistics management and the example of its application in the TC industry. In terms of an export market, competitiveness will involve the ability to select and deliver a set of product characteristics which matches the preferences in the importing country at minimum cost. In order to operationalise these elements of competitiveness for inclusion in a questionnaire, they need to be separated into discrete components. The logistics literature (Ballou, 1978) suggests the following components are relevant: the quality of communications between the importer and the export supplier; the percentage of orders delivered on time; the percentage of orders accurately filled and the incidence of various categories of error; the ease of monitoring the movement of goods from factory door to the purchaser's warehouse; the total time elapsed between the placing of orders and the delivery of the goods; and the ease with which incorrectly delivered or faulty goods can be returned.

The next section presents the results of a survey which converted these criteria into questions that were asked of Australian importers of TC products from Asia. Questions were also asked about some characteristics of the product at the factory door, price, quality of packaging and fashion content.

SURVEY OF AUSTRALIAN TC IMPORTERS

The survey was conducted to ascertain the extent to which individual aspects of logistics performance affect the competitiveness of Chinese TC exports in the Australian market compared to other Asian exporters in the Australian market.[3] This section contains detailed responses to survey questions and the results are summarised in the following section.

Attributes of TC at factory door ex-factory prices

Respondents were asked to consider the relative ex-factory price of an article imported from certain countries if the price for the article in China was 100. Only one respondent reported an ex-factory price less than 100 for one country (Pakistan). In all other instances, ex-factory prices in China were considered to be the lowest of all the Asian countries. The results suggest that China's prices are considered to be significantly less than those of most of the other Asian TC exporting countries with the exception of Pakistan (and Indonesia), whose prices are only marginally higher than those of China (see Table 7.3).

Table 7.3 Country Indexes of ex-factory price (China = 100)

	Mean	Standard deviation	Minimum	Maximum	Median	Number of respondents
Taiwan	131.7	18.4	110	180	120	19
Korea	126.6	14.0	110	160	120	16
Hong Kong	125.1	19.6	100	180	120	25
Philippines	120.0	(0)	120	120	120	1
India	114.5	15.8	100	150	110	12
Thailand	114.0	8.5	108	120	114	2
Pakistan	108.6	15.8	80	150	105	14
Indonesia	103.0	2.7	100	105	104	3

Packaging standards

For this question respondents were asked to take ideal packaging standards as 100 and rate all the countries against their ideal. Scores greater than 100 were occasionally awarded. Some firms which imported only raw cloth viewed this question as irrelevant and consequently did not answer it. In terms of packaging standards China ranks seventh, with only Pakistan and India perceived to have lower standards (see Table 7.4).

Fashion content

Respondents were asked whether fashion content was relevant in their business. Nineteen replied that it was. When these nineteen were asked to what extent the supplier provides the fashion input, their replies were: not at all (9 respondents), moderately (9 respondents), significantly (1 respondent) or completely (0 respondents). About half of the respondents provide all the fashion input themselves while the other half provide the majority of fashion input. Respondents were asked to score the fashion content of items from each country, taking leading European cities as 100. In general, only respondents who indicated that their suppliers provided some or all fashion input answered this question. In terms of fashion content, China, Pakistan and India all ranked significantly below other countries (see Table 7.5).

Table 7.4 Country indexes of packaging standards (ideal = 100)

	Mean	Standard deviation	Minimum	Maximum	Median	Number of respondents
Thailand	100	0	100	100	100	2
Hong Kong	95.3	11.4	60	120	100	24
Korea	93.7	11.1	70	110	100	15
Taiwan	92.1	11.3	60	110	90	17
Philippines	90.0	(0)	90	90	90	1
Indonesia	85.0	21.2	70	100	85	2
China	83.6	14.6	50	100	84	26
Pakistan	76.0	18.0	40	100	81	12
India	73.8	21.8	30	110	80	12

Table 7.5 Country indexes of fashion content (Europe = 100)

	Mean	Standard deviation	Minimum	Maximum	Median	Number of respondents
Thailand	90.0	0	90	90	90	2
Hong Kong	87.1	16.3	60	120	90	12
Indonesia	85.0	(0)	85	85	85	1
Taiwan	82.7	17.5	40	100	80	11
Korea	79.3	17.9	50	100	80	7
Philippines	70.0	(0)	70	70	70	1
China	56.3	24.8	20	90	60	12
Pakistan	55.0	26.5	20	80	60	4
India	51.7	27.9	20	90	55	6

Quality of communications

Overseas trips

Respondents were asked whether a buyer from their firm had ever visited any of the Asian countries. All thirty firms had sent buyers to China. Only one firm had not sent a buyer to Hong Kong (the firm was one which does not deal with Hong Kong). The majority of firms (22) had also sent buyers to Taiwan and Korea (21), while very few had been to Thailand (3), the Philippines (2) or Indonesia (3). About one-third of firms had sent buyers to India (13) and Pakistan (10).

Paperwork and bureaucracy

In this case, respondents were asked to rate the amount of paperwork in arranging for TC imports from the countries in question, taking 100 as an acceptable/reasonable amount of paperwork. A few respondents pointed out that the amount of paperwork for importing into Australia is fixed and does not vary between countries, and that the paperwork done on the exporting side is handled by the exporters and so does not affect the importers. Nonetheless, this question was answered by most respondents and their relative answers for each country are valid even though the interpretation of the absolute value is unclear. (That is, some scored all countries well below 100 and some scored above 100.)

There is not a great deal of difference in the means when the size of the standard deviations is considered. All medians were 100 except for Indonesia for which relatively little data are available. Although there is a wide variation in the data values, respondents tended to regard most countries as having a similar amount of paperwork and bureaucracy (see Table 7.6).

Table 7.6 Country indexes of paperwork requirements (reasonable = 100)

	Mean	Standard deviation	Minimum	Maximum	Median	Number of respondents
China	106.4	39.8	0	200	100	28
Thailand	100.0	0	100	100	100	2
Pakistan	95.4	28.8	30	130	100	12
Indonesia	95.0	7.1	90	100	95	2
India	94.2	34.2	10	140	100	13
Taiwan	86.3	29.0	15	110	100	19
Hong Kong	82.4	29.0	20	110	100	27
Korea	76.8	37.8	0	110	100	14

Effectiveness of telex and fax messages

Respondents were asked to estimate the percentage of telex or fax messages which were replied to promptly and accurately. It was suggested that interviewers consider 100 to be the level of promptness/accuracy that they consider ideal.

According to the respondents China, India and Pakistan reply least promptly and least accurately of all the countries. Hong Kong, Taiwan and Korea are perceived to be considerably more prompt and reliable in their communications (see Table 7.7).

Impact of language differences

Respondents were asked in what proportion of their dealings with the various countries were language differences a problem. Several respondents reported that they went to considerable lengths to avoid language differences and offered the view that their answers might not reflect the general experience of other importers.

Table 7.7 Country indexes of responses to questions (ideal = 100)

Promptly	Mean	Standard deviation	Minimum	Maximum	Median	Number of respondents
Hong Kong	94.6	6.7	80	100	98.5	20
Taiwan	93.6	7.9	80	100	97.5	16
Korea	91.3	8.3	80	100	92.5	12
Thailand	90.0	14.1	80	100	90	2
Philippines	80.0	(0)	80	80	80	1
India	71.5	24.7	40	100	80	10
Indonesia	70.0	42.4	40	100	70	2
Pakistan	69.6	23.3	30	100	80	11
China	64.9	26.0	0	100	70	22
Accurately						
Thailand	96.5	2.1	95	98	96.5	2
Hong Kong	92.7	11.7	60	100	100	21
Taiwan	91.8	12.2	60	100	100	17
Indonesia	90.0	0	90	90	90	2
Philippines	90.0	(0)	90	90	90	1
Korea	87.5	16.0	60	100	95	12
China	72.0	29.9	0	100	80	23
India	70.0	23.5	30	100	80	10
Pakistan	66.8	25.7	20	100	70	11

Of those countries for which sufficient data are available, India, Pakistan and China had the highest rates of language difficulties. The distributions of replies to this question are skewed because the majority reported no problems with any country. Although the level of problem was not reported to be great for the majority of respondents, China was the only country for which an adequate number of responses was received where the median proportion of communications hampered by language differences was not zero (see Table 7.8).

Percentage of orders delivered on time

Respondents were asked what proportion of consignments they would expect to be shipped after the date originally agreed. There was a large amount of variability in the answers to this question. China was perceived to ship the most consignments after the due date, closely followed by India and Pakistan. Korea, Hong Kong and Taiwan were all reported as shipping goods much more reliably (see Table 7.9).

Percentage of orders filled accurately, inspection and standard of goods on arrival

Thirteen respondents reported that goods were inspected prior to delivery. Eight of these also reported that inspection during delivery occurred. Respondents

Table 7.8 Proportion of dealings where language differences were a problem (per cent)

	Mean	Standard deviation	Minimum	Maximum	Median	Number of respondents
India	7.5	14.1	0	40	0	11
Pakistan	7.5	17.8	0	60	0	11
China	6.3	13.8	0	70	0.5	29
Korea	4.1	7.3	0	25	0	16
Taiwan	3.3	6.6	0	25	0	21
Philippines	2.0	(0)	2	2	2	1
Hong Kong	1.7	3.9	0	15	0	26
Indonesia	1.0	1.4	0	2	1	2
Thailand	0	(0)	0	0	0	1

Table 7.9 Proportion of shipments expected to be supplied after date agreed (per cent)

	Mean	Standard deviation	Minimum	Maximum	Median	Number of respondents
China	56.1	34.6	0	100	52.5	30
Pakistan	50.4	33.2	10	100	40	13
India	45.0	36.4	5	100	35	12
Philippines	30.0	(0)	30	30	30	1
Indonesia	27.5	31.8	5	50	27.5	2
Korea	17.7	24.6	0	100	10	17
Hong Kong	15.4	18.6	0	80	10	26
Taiwan	11.1	11.5	0	50	10	21
Thailand	10.0	0	10	10	10	2

were asked what percentage of products they would expect to be rejected at the time of delivery. In almost all cases the medians are substantially lower than the means, which is indicative of the presence of one or more very high rates of rejection for each country. Considering only countries where there is sufficient data to have any confidence in the ratings, China, India and Pakistan have the highest rates of rejection, while Hong Kong, Korea and Taiwan have lower levels (see Table 7.10).

Adjustments to orders

Firms were asked about the tolerance in each country to adjustments to orders after they have been finalised. This question was not well received because it seemed to imply that the firms were not well enough organised to avoid making

Table 7.10 Product rejection rates (per cent)

	Mean	Standard deviation	Minimum	Maximum	Median	Number of respondents
Philippines	10	(0)	10	10	10	1
China	8.6	16.4	0	80	2.0	29
India	7.5	14.7	0	50	2.0	11
Pakistan	5.7	5.1	0	15	5.0	12
Hong Kong	2.3	3.4	0	15	1.0	26
Korea	2.1	3.0	0	10	0.8	16
Taiwan	1.7	2.4	0	10	0.8	20
Thailand	1.3	1.1	0.5	2	1.25	2
Indonesia	0.8	0.35	0.5	1	0.75	2

late order changes. Many firms said that it was not an issue and that they did not make adjustments to orders after they had been finalised. Relatively few firms answered this question.

Korea, Taiwan and Hong Kong ranked as having the highest tolerance to adjustments to orders. China ranked fairly low but Pakistan was reported to be even less able to handle order changes (see Table 7.11).

Table 7.11 Tolerance to adjustments to orders (ideal = 100)

	Mean	Standard deviation	Minimum	Maximum	Median	Number of respondents
Korea	89.6	12.6	70	100	95	12
Taiwan	85.2	14.2	60	100	90	12
Hong Kong	80.5	27.9	0	100	90	15
India	73.1	32.8	15	100	85	8
Philippines	70.0	(0)	70	70	70	1
China	63.3	34.6	0	100	73.5	16
Indonesia	60.0	(0)	60	60	60	1
Pakistan	51.7	44.0	0	100	45	6

Time elapsed between ordering and delivery

Respondents were asked how many months ahead they would have to finalise an order for a new product. China ranks as the country with the longest lead-times, followed by India and Pakistan. (Indonesia and Thailand were also reported to require long lead-times but very few respondents answered for these countries) (see Table 7.12).

Table 7.12 Order lead-times (months)

	Mean	Standard deviation	Minimum	Maximum	Median	Number of respondents
China	8.7	4.6	3	24	7.3	30
Indonesia	7.5	(0)	7.5	7.5	7.5	1
Thailand	6.8	1.1	6	7.5	6.8	2
India	6.4	3.2	2	14	6.0	12
Pakistan	6.3	3.6	3	14	5.0	11
Korea	4.2	1.8	1.5	8	4.0	15
Taiwan	4.1	1.6	2	8	3.5	20
Hong Kong	3.6	1.4	1	6	3.0	25
Philippines	3.0	(0)	3	3	3.0	1

ANALYSIS OF RESULTS

Rescaling and ranking of attitudes

To overcome the dual problems of respondents having ranked different numbers of countries and having used different ranges to score the countries, a rescaling and ranking of their answers was undertaken. The medians of these ranks for each of the six countries for which there is a reasonable amount of data were then identified and have been tabulated (together with the means from the preceding text) for each of the twelve attitudinal questions (see Table 7.13). These rescaled scores can be regarded as a reliable and unbiased measure of the respondents' relative attitudes to the various countries.

The ranking of the medians of the rescaled scores is indicated in Table 7.13 by the figures in parentheses. These rankings have been determined such that a ranking of 1 indicates the 'best' country(s), while a ranking of 6 indicates the 'worst' country. China offers the lowest ex-factory prices of any country. On all other criteria China ranks worst or equal worst when compared to Hong Kong, Taiwan and Korea. Overall, China ranks similarly with India and Pakistan, since of the 12 characteristics considered, China ranks better than India and Pakistan for 3, the same for 4, and worse for 5.

A series of one-way analyses of variance were conducted on this data using country as the treatment. The results of these analyses are displayed in Figure 7.1. Each dot represents a country and the observation for China is marked by 'C'. Means which are not significantly different between the countries are underlined. For variables where higher scores were less desirable the signs of the means were reversed before they were plotted. Excluding ex-factory prices, China's mean was the lowest for 8 of the remaining 11 variables. In the remaining cases, China's mean was not significantly different from the mean of the country with the lowest score (except in the case of handling costs).

Table 7.13 Means of raw scores ($\bar{X}$) and medians of rescaled scores (SM) for each of the six main countries[a]

	China		Hong Kong		Taiwan		Korea		India		Pakistan	
	$\bar{X}$	SM	$\bar{X}$	SM	$\bar{X}$	SM	$\bar{X}$	SM	$\bar{X}$	SM	$\bar{X}$	SM
Prices[b]	(100)	(1)	125	0.59 (4)	131	0.67 (5)	127	0.50 (3)	115	0.50 (3)	109	0.17 (2)
Packaging	84	0.25 (3)	95	0.75 (1)	92	0.70 (2)	94	0.75 (1)	74	0.11 (5)	76	0.19 (4)
Fashion content	56	0.10 (5)	87	0.87 (1)	83	0.70 (2)	79	0.67 (3)	52	0.14 (4)	55	0.04 (6)
Finalise order (mths)[b]	8.7	1.0 (5)	3.6	0.25 (1)	4.1	0.33 (2)	4.2	0.33 (2)	6.4	0.64 (4)	6.3	0.50 (3)
Paperwork & bureaucracy[b]	106	0.50 (2)	82	0.50 (2)	86	0.50 (2)	77	0.37 (1)	94	0.50 (2)	95	0.50 (2)
Product rejection[b]	8.6	0.50 (2)	2.3	0.33 (1)	1.7	0.33 (1)	2.1	0.50 (2)	7.5	0.67 (3)	57	0.87 (4)
Messages prompt	65	0.0 (4)	95	0.75 (1)	94	0.75 (1)	91	0.75 (1)	72	0.37 (2)	70	0.17 (4)
Messages accurate	72	0.25 (5)	93	0.69 (2)	92	0.70 (1)	88	0.50 (3)	70	0.27 (4)	67	0.25 (5)
Language difficulties[b]	6.3	0.50 (1)	1.7	0.50 (1)	3.3	0.50 (1)	4.1	0.50 (1)	7.5	0.50 (1)	7.5	0.50 (1)
Late shipping[b]	56	1.0 (6)	15	0.28 (1)	11	0.30 (2)	18	0.33 (3)	45	0.67 (4)	50	0.80 (5)
Order adjustment	63	0.15 (4)	81	0.67 (1)	85	0.64 (2)	90	0.64 (2)	73	0.50 (3)	52	0.50 (3)

Notes: a Ranking of medians is indicated in parentheses.
b Lower scores are more desirable and have been ranked higher.

Analysis relative to principal type of imports

As the production characteristics of textiles and clothing are quite different, respondents' attitudes were examined separately for the two types of imports. Each respondent was assigned to one of three categories: import mostly textiles (>70 per cent textiles), import mostly clothing (>70 per cent clothing), or import equal amounts of textiles and clothing. As only one respondent fell into the last category the data for that firm were excluded from this analysis. Sixteen respondents imported mostly textiles while 13 imported mostly clothing.

In the case of China, those who import clothing rather than textiles have relatively more problems with packaging, fashion content, rejection of goods at

Figure 7.1 Means of rescaled scores showing significant differences between the countries

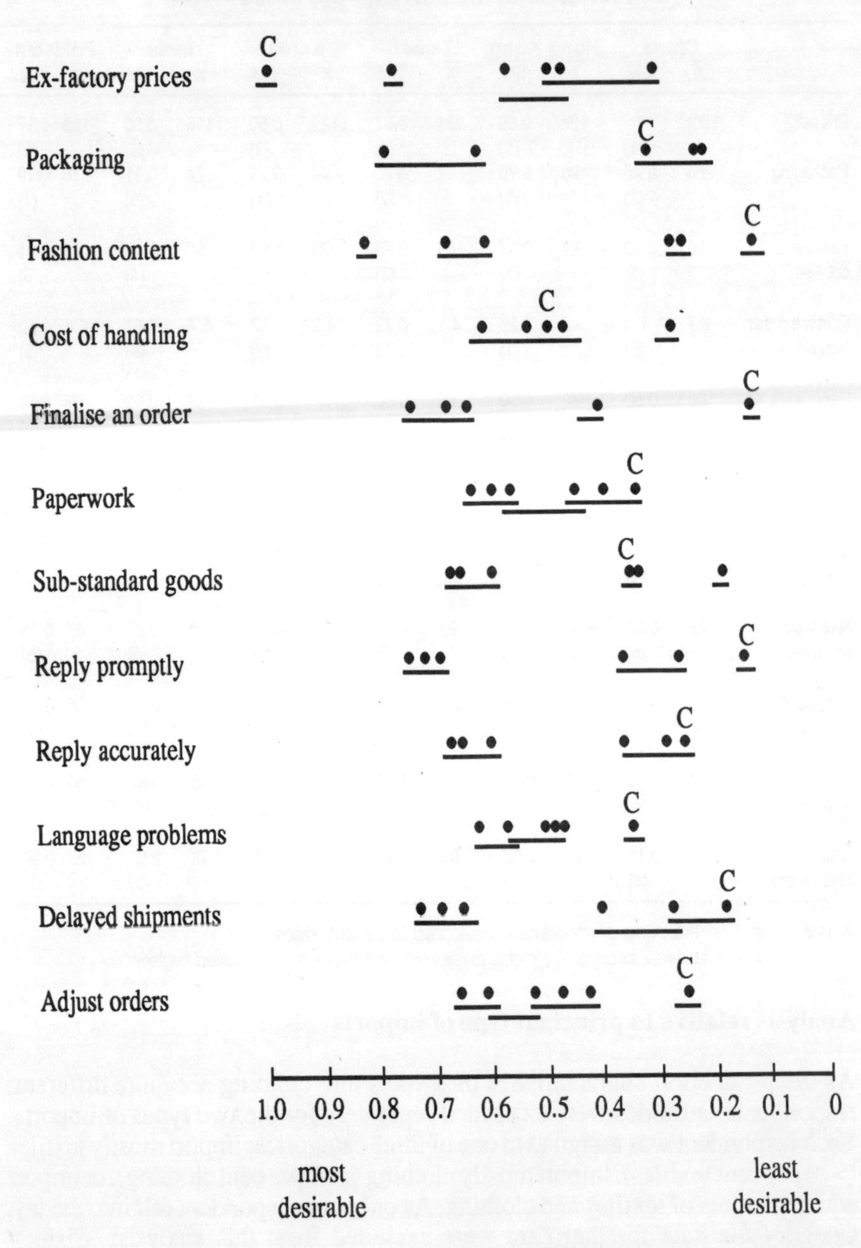

Note: Countries joined by a line have means which are not significantly different.

delivery, accuracy of messages and tolerance to order adjustments. The costs of handling are higher and there is more paperwork and bureaucracy. Importers of clothing do however tend to have less trouble with language differences. Considering the differing requirements for the production of textiles and clothing, the nature of these comparisons is not surprising.

The significance of apparent differences in results between types of imports was tested in a series of one-way analyses of variance using textiles/clothing as the treatment. There were, however, no significant differences between the means of the scores for the two groups.[4]

Analysis relative to size

The majority of respondents answered the question about the value of their annual imports. Those who did not wish to give a numerical value indicated whether they were small, medium or large-sized firms. For the purposes of this analysis small firms are those with imports valued less than $A1m, medium-sized firms are those with imports valued between $A1m and $A8.9m, while large firms imported at least $A9m in value annually.

In the case of China, small firms tended to fare worst with respect to packaging and fashion content. Medium-sized firms appeared to fare worst with costs of handling, accuracy of replies, language difficulties, late shipments and tolerance to order adjustments. Large-sized firms tended to fare worst with paperwork and bureaucracy and proportion of product rejected.

As a result of observing these tendencies, a series of one-way analyses of variance using size as the treatment was conducted. However, these resulted in no significant differences between the means of the scores for the three size groups when taken across all countries. Either the measure of size is not very good, or there is no consistent relationship between size and attitudes/experience as an importer of textiles and clothing. When a series of two-way analyses of variance was conducted on size and country, country was highly significant in all cases but size was not.

Summary

While China has the lowest factory price of any country covered in the sample, it ranks lowest (or equally low) on other criteria such as packaging, fashion content, the time to finalise an order, replies to messages, language problems, delayed shipments and the flexibility in responding to adjustments to orders. The responses do not differ significantly between textiles and clothing importers nor between importing firms of different sizes.

It was argued above that these characteristics would influence the competitiveness of the exporting country in import markets. In other words, we would expect to see a correlation between the performance of various exporting countries according to these criteria and their penetration of export markets. Also, as explained above, it is possible to measure these correlations in the Australian import market for textiles and clothing, and the results are summa-

Table 7.14 Partial correlation coefficients between import market shares and survey responses

Characteristic	Textiles		Clothing	
	1980	1986	1980	1986
Packaging	+++	+++	+++	+++
Fashion content	++	++	++	++
Finalise order				
Reject (per cent)	---	---	---	---
Paperwork				
Prompt reply	+++	++	+	
Accurate reply	+++	++	++	
Language problems				
Late shipments	--	-		
Change orders				

Note: Correlations between survey respondents' assessments of performance and Australian import market shares. The signs indicate the direction of correlation. The number of signs indicates significance (1 represents 5 per cent, 2 represent 1 per cent, 3 represent 0.1 per cent).

rised in Table 7.14. Price is obviously an important determinant of market penetration. However, our question here is, why, if China is such a competitive supplier at factory door, did its penetration of the Australian market not grow faster? Our hypothesis is that factors other than price have offset China's overall competitiveness. We therefore examined the correlations between Australian import market shares and the survey variables for both textiles and clothing across all countries, and the influence of price was removed by calculating partial correlation coefficients. The results show that the significant variables were packaging, fashion content, replies to messages, the product rejection rate and, to a lesser extent, late shipment. Generally, the results did not differ markedly between textiles and clothing.[5] The implications of these results are explored in the next section.

IMPLICATIONS

Logistics performance matters

The results support our contention that the logistics performance of exporters, even of apparently simple products such as textiles and clothing, does matter when it comes to penetrating foreign markets. However, the results also indicate that not all logistics characteristics are equally important. The characteristics which appear to matter most are those which become critical once the order is fixed. The buyer can allow for language difficulties, and for the time taken to complete an order. With some care, it will be possible to avoid having to change orders at later dates, thereby avoiding the problems of the lack of flexibility of the suppliers. The characteristics which matter most are those which are not amenable to offsetting strategies or which are outside the control of the buyer. These include, in particular, the activities which must occur after the order is made. They include packaging, quality control, shipment times and responses to communications. The same applies to fashion content when the buyer relies on the seller.

The implication for China is that improved performance in terms of these characteristics of service once the order is placed will improve market penetration. An improvement will permit China to capitalise on its highly competitive position in terms of price at the factory door.

The survey results apply to exports to a high-income country but the degree of importance of the characteristics identified may vary between export destinations. In particular, the importance of these logistics variables may offset to some extent the otherwise high degree of complementarity between countries at different stages of development (as discussed in Chapter 5). Industrialised countries may put high weight on logistics characteristics which their developing country suppliers may find relatively expensive to supply. The reason is that the activities necessary to supply the service characteristics may be relatively capital or technology-intensive, unlike the process of production at the factory level. Thus import-competing firms in industrialised countries may be able to maintain market share if they can achieve a higher level of logistics service quality at lower cost than foreign suppliers.[6]

Logistics services are internationally tradable

The argument of the previous section that the supply of logistics services may differ in its factor intensity to the basic production process at the factory level raises the possibility of extensive international trade in these services. In the case of China this has already been observed. Sung (1988) has examined in detail the role of Hong Kong in supplying these sorts of services to China. The supply mechanism is that Hong Kong imports the goods from China then re-exports[7] them with various other logistics services added to them. Sung estimates that in 1984, re-exports from Hong Kong accounted for 21 per cent of total Chinese exports of textiles and 28 per cent of clothing exports.

Hong Kong thus plays an important role for China in acting as an intermediary between China and the industrialised country markets which have substantial logistics requirements. However, in this case, the Multifibre Arrangement (MFA) takes on a new significance. The MFA limits imports by country quota and it imposes strict rules on 'country of origin'. The MFA therefore restricts not only trade in goods which are directly subject to its regulations but also affects the trade in logistics services. Because of the export quotas and the country of origin rules, China either must import less logistics services from Hong Kong and therefore export less textiles and clothing than would otherwise be the case, or else it must seek other suppliers of those services who are not limited by MFA rules. It is not clear, however, that these other suppliers will be as competitive as Hong Kong from the Chinese point of view. In other words, China may have to pay a higher marketing margin to suppliers other than Hong Kong.

Reform of the management of China's foreign trade

A further implication of restrictions on international trade in logistics services, as a result of restriction on goods trade as noted above, is that there is extra pressure on Chinese suppliers to become self-reliant in the supply of these services. The incentives to respond to that pressure are limited by problems in the system of the management of foreign trade.

These problems were reviewed by the World Bank (1988b, ch. 2) in its recent China country study (and referred to in Chapter 3 in this volume). The World Bank study discusses the problem of the air-lock between Chinese enterprises and world market forces. It is argued that the institutional arrangements, particularly the role of the foreign trade corporations and the methods of pricing goods for export, has insulated enterprises involved in production from the world market-place. Thus it will also have isolated them from the incentive to deal with the logistics issues which are highlighted in this chapter.

In the last few years there has been a trend towards much greater decentralisation in China's foreign trade system. However, in 1989 the macroeconomic problems and the lack of market mechanisms to deal with those problems in China has led to a reinstatement of central control over the trade system, including the withdrawal of some rights to deal directly with the international market.

Infrastructure issues

Attempts to improve the logistics performance of the Chinese export sector will inevitably focus attention on issues related to the transport and communications infrastructure in China. The productivity of China's transport system is low compared to other countries.[8] There are expected to be positive returns to investment in the transport and communications infrastructure. Investment strategies could be classified into four categories:

1. measures to improve the performance of individual sections of the transport system, thereby reducing costs and uncertainty;
2. measures to improve the connections between sections of transport and handling system, such as larger container sizes which reduce handling costs, or larger warehouses;
3. measures to reduce the overall time-span of export transactions, for example, greater capacity in the communications system; and
4. measures which reduce the probability of malfunction in the system.

However, over the time-scale relevant to assessing China's prospects in penetrating foreign markets, these sorts of investments are not likely to be significant.

More important in the shorter run will be the incentives of local exporters to work in and around the existing infrastructure. The demands for services that they create will also focus attention on the shortcomings of the infrastructure and make it more likely that the sorts of strategies listed above will be pursued. According to this argument, the institutional arrangements for the management of international transactions, and the incentives that are created, will become increasingly important determinants of China's logistics performance both now and in the longer term.

8 Collective resource management in China: the raw wool industry

LIU ZHENG, CHRISTOPHER FINDLAY AND
ANDREW WATSON

INTRODUCTION

Since the rural reforms began in 1978, the use and management of collective assets in the Chinese countryside has emerged as a major issue. In many cases, assets which can be controlled effectively at the household level are now being managed more efficiently. However, household autonomy and the withdrawal of local collective organisations from some of their previous responsibilities has led to concerns about the running down of those assets which are collectively owned but privately used. This concern is often expressed in relation to the physical infrastructure of a community, such as its irrigation system or transport facilities. This chapter reviews this issue in relation to a key collective asset involved in raw wool production, namely pasture land. In the Chinese setting, the passing down of decision-making power has created problems in the maintenance of the value of the asset. These problems have, in turn, influenced the growth of raw wool production.

One analytical structure adopted in this chapter in order to identify and analyse this problem is that of 'the tragedy of the commons' (Hardin, 1968), in which the uncontrolled access by individuals to a jointly consumed resource, in circumstances where the users are not required to take into account the impact of their use on others, leads to the denigration of that resource. This analytical approach is entirely appropriate for the period since the reforms of 1978. The shift to the household responsibility system was a response to the inadequacies of the planning system, but, because of the common property nature of the pastures, that policy change has reduced, not enhanced, the productivity of land and labour in the sheep industry.

The deterioration of China's pasture land, however, has a much longer history than that brought about by recent policy changes. We therefore also argue that the problem of the pastures illustrates the longer run tension in China involved in the interaction between settled agriculture practised by the Chinese and nomadic herding practised by the other nationalities of north and central Asia (Lattimore, 1962). Over a very long historical period, as the area of settled Chinese agriculture expanded, pastoralists were forced out towards the more marginal areas. Although our focus in this chapter is the period since 1949, the same forces in favour of cultivation have been evident. This tension has also been exacerbated by the demands of the planning system and the transfers of resources in favour of urban residents which it involved. We give instances below, for example, of the pressures from the centre to raise stock numbers despite the deteriorating quality of the pasture. Even more recently, the tension emerges, in ways which we explain, in the choices over the types of sheep which should be run in China's pastoral zones. Both these examples can be seen as purely economic issues. At another level, however, they represent the latest phase in a long process of cultural interaction.

In summary, this chapter contains two analytical themes: one, the longer run process of the conflict over the distribution of the resources and wealth between the Chinese and the nomadic nationalities; and two, the issue of the management of the common property resources, such as pasture land.

This chapter is also a vehicle, in terms of the structure of this volume, for discussing and evaluating some of the technical issues involved in the production of wool in China. Average raw wool yields, for example, are relatively low in China, of the order of 2 kg per sheep, compared to a yield of double that and more for Australian merinos. What, then, are the factors constraining the level of yields and the growth in Chinese raw wool output? The answers can be divided into two groups. One focuses on the technical determinants of wool production, especially the genetic characteristics of the Chinese sheep flock, and the attributes of the grasslands. The other focuses on the managerial and incentive consequences of the economic reforms. In the following sections of this chapter, we first examine factors in China's natural pastoral environment which shape raw wool production and then proceed to discuss grassland management, genetic factors and the provision of rural infrastructure. All of these issues illustrate the underlying themes raised at the outset of the relationship between private producers and collective assets, and the longer run forces towards the degradation of pastures in the process of Chinese development. The other managerial, incentive and pricing issues are discussed in Chapter 9.

WOOL PRODUCTION REGIONS IN CHINA

Animal husbandry districts

As a result of the pattern of physical geography and historical development, most of China's sheep are located in the arid and semi-arid pastoral areas stretching across China from the northeast to the northwest, and they are largely raised by members of China's national minorities. Only a few sheep are grazed in the farming areas of central and southern China. Seen in these terms, therefore, the raising of sheep is not only related to environmental issues and the conflict for land use between crop farming and animal husbandry, it also reflects important

Map 8.1 Agricultural districts in China

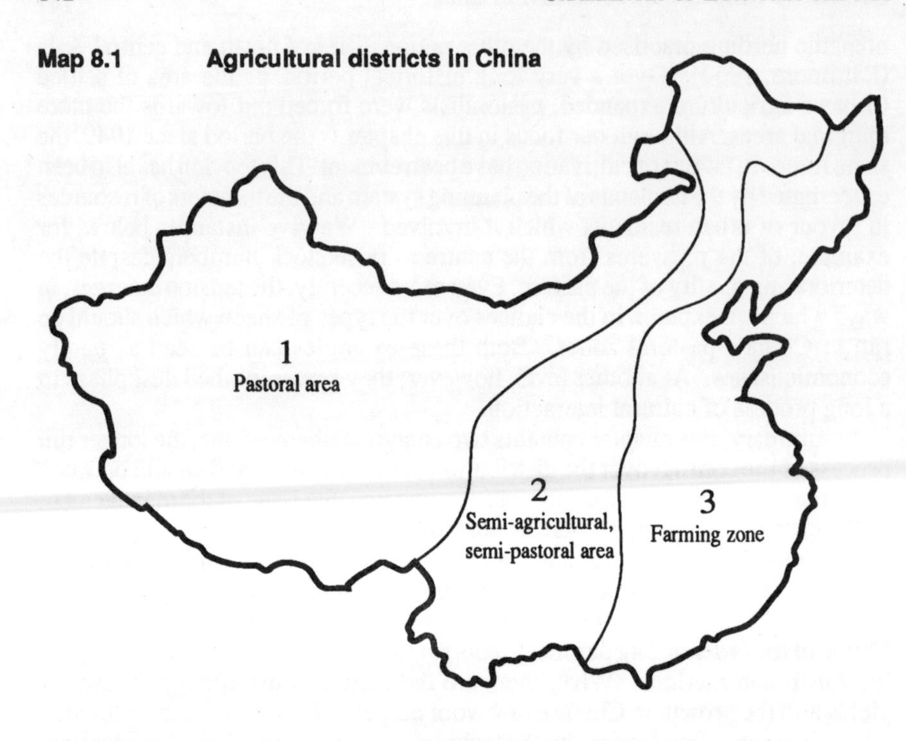

Map 8.2 Animal husbandry districts in China

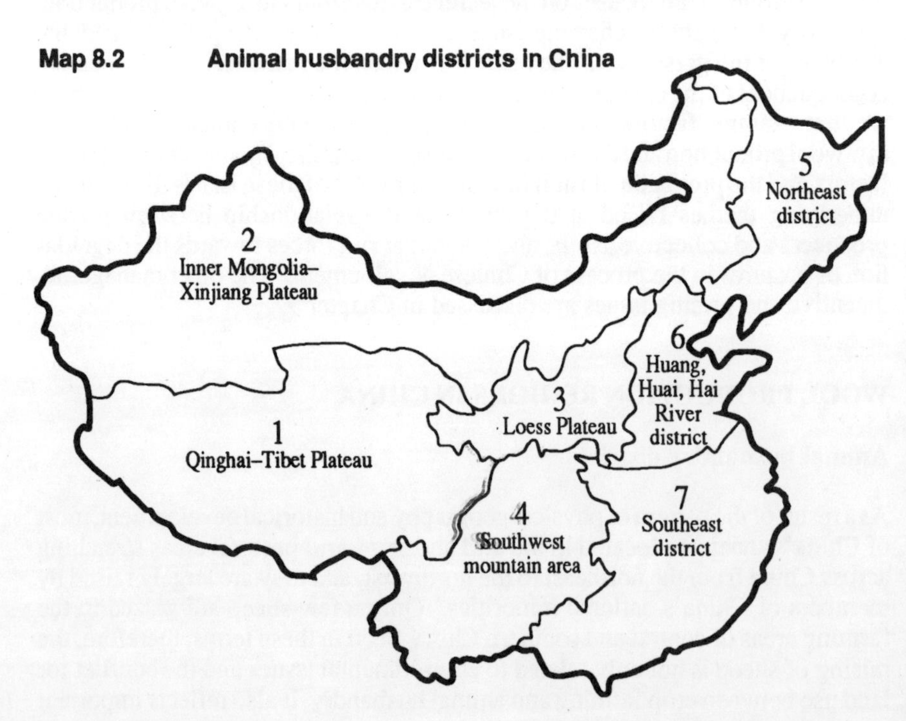

cultural differences between people with contrasting ways of life (Lattimore, 1962).

At the broadest level, China's countryside can be divided into three production zones: the pastoral zone; the semi-pastoral, semi-agricultural intermediate zone; and the farming zone (see Map 8.1). Roughly speaking, Inner Mongolia–Xinjiang and the Qinghai–Tibet plateaus comprise the pastoral zone; the Loess plateau and the southwest mountain district make up the semi-agricultural, semi-pastoral zone; and the other coastal districts make up the farming zone. These broad zones correspond to the general features of rainfall, temperature and environment. At a more specific level, however, the animal husbandry districts of China (including those in the non-pastoral regions) can be further identified by pasture type and productivity, the nature of the terrain and the characteristics of the animals raised. Map 8.2 shows these animal husbandry districts with much finer detail than the broad zones as typically defined. These are the districts which will be used to organise the data in the rest of this chapter.[1]

Most Chinese sheep are raised in those animal husbandry districts which lie within the broad pastoral zone, particularly the Inner Mongolia–Xinjiang and Qinghai–Tibet plateaus, which in 1987 had 71.3 per cent of China's total sheep numbers (see Maps 8.3 and 8.4). The Inner Mongolia–Xinjiang plateau is the main wool-producing region of China. Finewool and semi-finewool sheep account for 47.7 per cent of sheep in this region (the average proportion for the whole country is 37.5 per cent). The local coarsewool sheep (mainly Mongolian and Kazakh sheep) account for 52.3 per cent. The latter are bred for both their meat and their wool, and the local national minorities have a preference for the meat of these breeds. On the Qinghai–Tibet plateau, where the weather is severe and cold, the main breeds are coarsewool Tibetan sheep, used for carpet-making. These make up 89 per cent of the sheep in the district.

The Loess plateau and the southwest mountain district form the intermediate zone between the pastoral and agricultural areas. The main breeds of sheep in the Loess plateau are the Tan sheep, which are bred for lambskins, and other local sheep bred for meat and wool. Goats also account for a big proportion of total flock numbers. In the southwest mountain district, the proportion of goats is around 68 per cent of the total flock. Sheep here mainly produce carpet wool.

Cross-farming predominates in the remaining districts, and sheep-raising is much less important for the rural economy. In the northeast district, grain is the main agricultural product and the number of sheep make up only 5 per cent of the national total. More than 88 per cent of these, however, are finewool and semi- finewool sheep. The Huang-Huai-Hai district is mainly a grain and cotton production area. Crop by-products are the main source of feed for sheep-raising. The number of sheep in this district account for nearly 10 per cent of the national total and of these some 50 per cent are finewool, semi-finewool and improved breeds. The Han sheep is the predominant breed in this area, and it produces high quality carpet wool. This district is also a major area for the production of goat skins and milking goats. Goats account for 70 per cent of the total flock here and nearly one-third of China's total goat numbers. The southeast district is the main rice production area. Its grain production accounts for 40 per cent of national output. Livestock include buffalo (60 per cent of the national total), pigs (40 per

Map 8.3 **Chinese sheep and goat numbers (million head) by region, 1986**

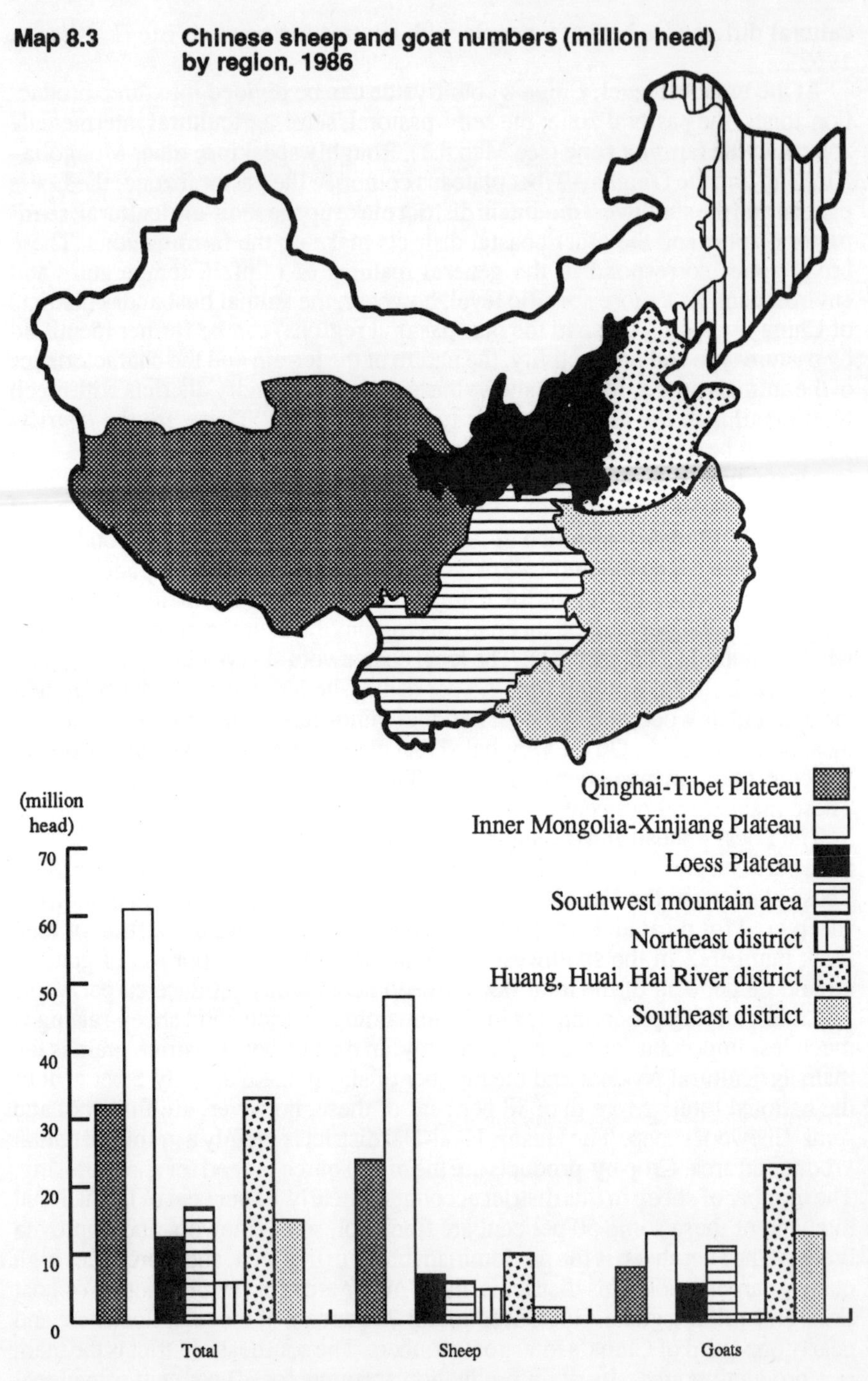

Source: *Agricultural Yearbook of China*, 1988.

Map 8.4 Distribution of sheep numbers by region, 1987 (per cent)

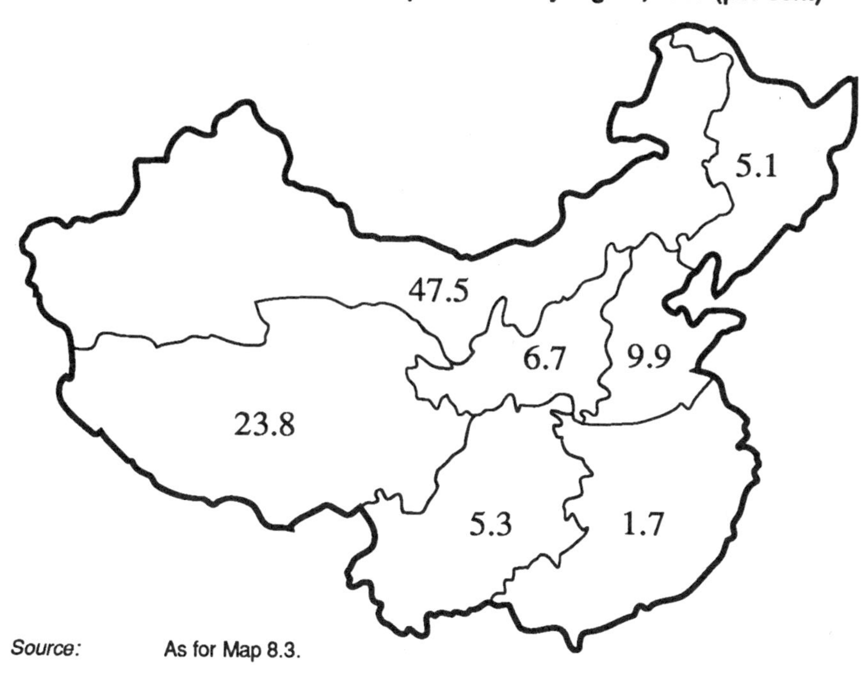

Source: As for Map 8.3.

cent) and ducks. There are more than 330 million mu (1 hectare = 15 mu) of hilly and high grasslands in this region, but because the climate is tropical and subtropical, with high rainfall and high temperatures in summer, it is unsuitable for sheep and cattle. Also, the mountains in southern China are very steep and it is very difficult to graze sheep and cattle naturally. As a result, the number of sheep in this region is less than 2 per cent of the national total.

Climate and vegetation

Herbage yield is dependent on topography, soil and climate, as well as on the natural characteristics of the grasses which make up the pasture and the state of their utilisation. Of these various elements, the essential one is climate, which affects output through moisture and heat. Differences in the combination of moisture and heat throughout the year also mean that seasonal variations in grass growth are very marked. In north and northwest China hot summers with a relatively higher rainfall followed by severe cold and dry winters lead to a pattern of good herbage growth in summer and autumn and withered and yellow pastures in winter and spring. This alternation is reflected in the quality of the sheep flock and in wool output. Numbers decline in winter and spring, and the tensile strength of the wool fibre is weakened by nutritional shortages which lead to break points in the wool. The precise features of these climatic conditions for each of the animal husbandry districts are summarised by the 'climate index' reported in

Map 8.5 Climate characteristics of animal husbandry regions

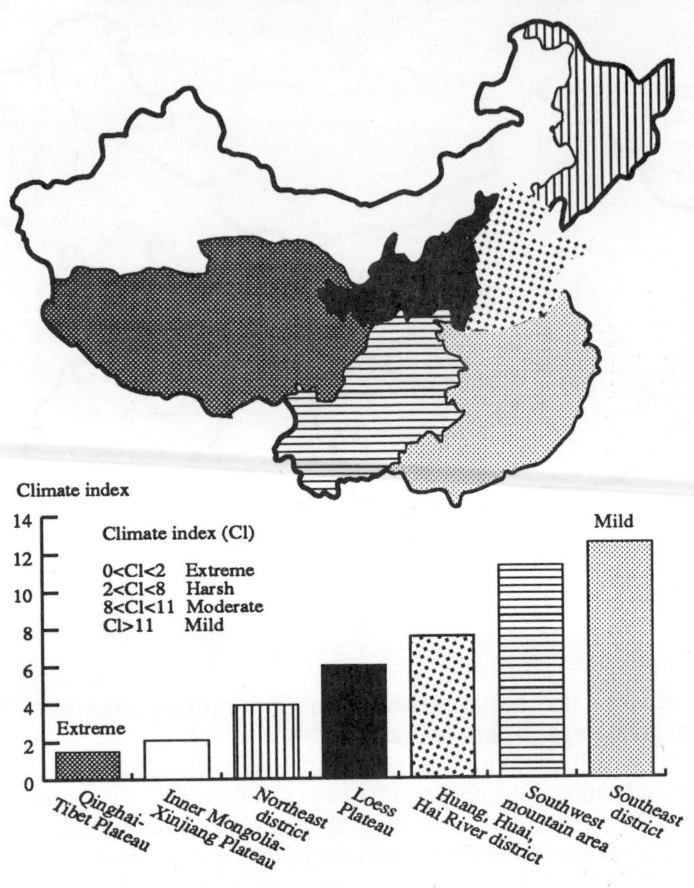

Note: The climate index is based on relative provincial ranking of temperature, rainfall and frost-free days. Climatic data are the average of each region. Within provinces, data are for the harshest area and, where data are not available, are estimated for an equivalent neighbouring area. These figures should only be taken as an approximate indication.

Source: State Statistical Bureau, China, *China Urban Statistics*, 1985.

Map 8.5. This index is based on average annual temperature, rainfall and the number of frost-free days. Low values of all these variables create very harsh conditions for sheep-raising. According to this index, the climate is harshest on the Qinghai–Tibet plateau, where altitude intensifies the general features shared by the pastoral areas.

In general, the northern and western areas of China have sparse rainfall and are typical arid and semi-arid zones (see Map 8.5).[2] Underground water is at great depth, there is little surface water and the rivers are intermittent. The quantity of rainfall is thus the major factor determining the yield of herbage. Except for its eastern end, most of the pastoral areas have precipitation of less than 300 mm annually, and some have below 100 mm. Some 60 to 70 per cent of this precipitation occurs in the summer, 15 to 20 per cent in autumn and about 15 per cent in winter. There are also quite marked annual variations. Generally speaking, precipitation in the wettest year is twice as much as in the driest, and the difference can approach a factor of four.

The development of animal husbandry and agriculture in the pastoral zone is thus strongly constrained by this aridity. Dry crops can be planted without irrigation in the semi-arid areas but the harvest is unstable, and the soil is susceptible to erosion. In the arid areas, except where ground water or snow water from the mountains is available, most of the land is not irrigated. The cultivated area of the zone is less than 10 per cent of the total cultivated area of the country as a whole. Much of the remaining land is desert and animal husbandry can only develop where there is sufficient pasture and water for stock.

Seasonal variations in temperature are also very large in the pastoral areas. In general, the cold is severe and the frost-free period is short. Winter is the driest period and December, January and February are the coldest months. The summers are moister and the warmest months are June, July and August, when most of the rain falls. Most grasses will grow when the average daily temperature rises above 5° centigrade and when there is adequate moisture, though some varieties need a higher temperature and are slower to become verdant (Science Publishing House, 1977). In northwest China, therefore, pastures are best during summer, and the grass withers and yellows after the onset of winter. Apart from any physiological characteristics, the important factors are the lack of moisture in the soil and the low temperature. The change is often not gradual but occurs rapidly after a sudden drop in temperature. Output of grass is also related to the quantity of snowfall in winter. If there is heavy snow, the spring thaw increases the soil moisture and provides a good basis for the growth of grass. Little snow in winter results in poor soil moisture and delays in the growth of grass.

Because of these seasonal constraints, mating generally takes place in autumn in order to ensure that the lambs are born in spring (March to April). They then have a warm climate with sufficient green pasture to promote growth and do not have to face the severe cold of winter. Shearing occurs in mid to late spring so that by June the clip is in the wool stores. The sheep are grazed on natural pastures throughout the summer and into autumn or early winter. They are then moved to winter pastures or sheds. In some mountainous regions, such as Xinjiang, the pattern involves the movement of flocks between winter, spring-and-autumn and summer pastures. Lambing and shearing take place on the spring-and-autumn pastures. The flocks then move to the high summer pastures for summer grazing. They return to the spring-and-autumn pastures for mating, slaughtering and flock management. Finally, they move to winter valleys and graze on special winter pastures supplemented with stored hay and concentrates, or out into oases on the fringes of the large desert basins. In other areas, the sheep are hand-fed in sheds in winter using stored hay, crop residues, and soya bean cakes. This winter period of hand-feeding is long. During this time, feed shortages often mean that the animals stop growing and their body weight declines sharply. They may also be required to move a long distance to look for feed and, as a result, most of the nutrients absorbed cannot be use for growth but are lost in the energy used for migration. Commonly, neither shed space nor stored feed is sufficient, and the sheep have to be grazed on natural pasture for the whole year.

Natural calamities

Because of the low rainfall and the large annual variations, drought is one of the major problems facing the sheep industry in pastoral areas.[3] Dry seasons entail two main problems. When the spring is dry, the period of lush growth is delayed. If there is little rainfall in summer, the grass grows poorly, and the output and quality of the herbage declines, which results in a shortage of feedstuff for the stock, especially for the following winter. The yield may be as low as 25 per cent of that of a wet year (Science Publishing House, 1977).

'Black disaster' is the term most commonly used in Inner Mongolia to describe the problems caused by the lack of snowfall in winter (Science Publishing House, 1977). Some areas depend on winter snows for stock drinking water. When there is little or no snow, there are heavy losses. Lack of early snow means that the movement of sheep to their winter pastures may have to be postponed. Lack of late snow means that they may have to be withdrawn ahead of time. With a shortage of water, the sheep inevitably lose weight and weaken. In poor condition they are easily infected, and disease in the sheep flocks can then cause enormous losses. In addition, nutritional problems cause slow growth and break points in the wool fibres.

By contrast, 'white disaster' refers to too much snowfall. The snow is too deep and affects normal grazing activity. When the whole ground is heavily covered, the sheep have insufficient or no feed, and this again results in thin and weak animals, miscarriage of lambs, a decline in the survival rate of newborn lambs and an increase in the death rate of old and weak animals. In normal years, the death rate of sheep is about 3 to 5 per cent (including death from diseases and wolf attacks) but during a 'white disaster' period the death rate commonly rises to 8 to 10 per cent generally and may even approach 40 to 50 per cent.

DEGRADATION OF THE GRASSLANDS

Degradation of the grassland is one of the most important factors limiting the development of the sheep industry in the pastoral areas. The main reasons for degradation are increased cultivation, deforestation and overgrazing. Degradation results in soil erosion, salinisation, an increase in the growth of toxic and unwanted plants, and a decrease in the output, quality and variety of edible herbage.

Increased cultivation

As a result of economic development since 1949, a large amount of previously cultivated land has been taken over for other uses. In the last four decades, cultivated land has been lost at an average of more than 7.8 million mu per year. The total area lost over the period is estimated to be about 300 million mu (Wu

Jinghua, 1988). Despite this loss, however, the cultivated area has remained about the same as in the initial stage of development after 1949. The reason is that some 350 million mu of former pasture land and 80 million mu of forest were opened up for cultivation. Since agriculture inevitably spread into regions with fertile soil and a relatively good climate, most of the best pasture was thereby lost to crop-farming.

In the region along the Songhuan and Nen Rivers in northeast China, for example, there were about 60 million mu of pasture in 1949. Production of grain became a key target and, with extensive cultivation, by 1980 there were only 36 million mu of prairie left — an average decrease of 760 thousand mu a year (Song et al., 1988). Because of this increased cultivation, the natural vegetation of the grassland was damaged, and erosion by water and wind was aggravated. This not only harmed animal husbandry but also weakened the ecological basis for agricultural production.

In Inner Mongolia more than 37 million mu of high quality pasture land were brought under cultivation during the Great Leap Forward and the Cultural Revolution (Qian Fenyong, 1987). The ecological balance was disturbed, and sand erosion and degradation now affects more than 400 million mu — one-third of the total grassland of the whole autonomous region. Furthermore, most of the newly cultivated land was not used continuously for crop production. Some of it was only ploughed for one or two years. As a result, top quality pasture became poor farming land. Eventually, because the prairie's primary vegetation cover was destroyed, this newly cultivated land was engulfed by sand-hills.

In sum, the cultivation of pasture land pushed agriculture into areas that were too marginal to survive. The result was both a decline in pasture area and a decline in total pasture quality. Furthermore, this process occurred over a period when there was a large increase in total stocking rates.

Deforestation

Deforestation has occurred through the removal both of individual trees and of great tracts of forest. A further problem is that, due to a lack of fuel for cooking, herders have felled or cut the bush scrub in the pastoral areas, resulting in much damage to the grassland from sand and wind. Originally, there were more than 6 million mu of desert poplar forest in the oases of the Talimu Basin in Xinjiang. That has now decreased by 3.4 million mu. Of 35 million mu of red willow, just 15 million may survive (Wu Jinghua, 1988). Similar examples can be cited from many other areas.[4] Deforestation is thus the result both of turning land over to cultivation and of human demand for fuel. Inevitably, the removal of trees and scrub reduces resistance to wind erosion and changes the local climate by lowering moisture retention.

Overgrazing

Since 1949 the concentration on sheep numbers and the unplanned use of pastures has resulted in a long-term trend towards overstocking. This has happened despite the overall reduction in pasture area described above. National sheep and goat flocks grew from 26.22 and 16.13 million respectively in 1949 to 110.57 and 90.96 million in 1988. Even without a reduction in pasture area, stocking rates would have inevitably increased. Furthermore, sheep are not the only animals grazed on the pastures, and large increases have also occurred with other animals. Total Xinjiang animal flocks, for example, grew from 10.4 million in 1949 to 30.2 million in 1983 (Li Yuxiang, 1986: 30, 35). In the early 1950s Xianghuang Banner in Inner Mongolia had little stock and good grassland, with lots of potential to develop its sheep industry. In 1949 there were about 100,000 head of stock, equivalent to an average load of 2.7 sheep units per 100 mu. By the end of 1982 stock numbers approached 3.3 million head, equivalent to 7.9 sheep units per 100 mu. The stock density was three times as much as that of twenty-three years before (Ai Yunhang, 1987). The grassland was over-grazed and the natural productivity gradually decreased.

Pasture land was under considerable pressure under the collective system also. According to one report, in Inner Mongolia the amount invested per unit area in pasture improvement over the past thirty years has been less than one-seventieth the value of animal husbandry products per unit area over the same period. As a result, pasture yields in some areas are now less than half of what they were in the 1960s (China, State Council Rural Development Research Centre, 1987: 49). Although, as discussed below, the introduction of the household responsibility system has in many ways exacerbated this problem, it is important to recognise that this is part of a long-term trend in China's animal husbandry.

Effects on grassland productivity

These problems have had severe effects on the productivity of the grasslands, which has decreased on average by 30 per cent from 125 kg per mu in the 1950s to 85 kg per mu in the 1980s (Wu Jinghua, 1988). The theoretical stocking capacity of the natural grassland has also declined year by year.[5] For example, in the 1950s the Inner Mongolia prairie had a capacity of 87 million sheep equivalent units. In the 1960s it dropped to 85 million units and to 65 million units in the 1970s. In the early 1980s it only had a capacity for 58 million units (Wu Jinghua, 1988). According to one estimate, among the 266 stock raising, semi-farming and semi-pastoral counties in China, 87 are overgrazed, 106 are approaching saturation point and only about 73 have any potential for increasing flocks (Wu Jinghua, 1988).

A similar process has taken place in Xinjiang. According to one source (Li Yuxiang, 1986: 164–5), by the mid-1980s more than one-third of Xinjiang's 700 million mu of pastureland was degraded, and herbage yield had generally declined by 40 per cent and more over the previous 40 years. At the same time, the problem of the increasing growth of toxic plants, caused by animals overgrazing the good grasses and leaving only the toxic ones to spread, has worsened. One of the causes of this change was the cultivation of some 60 million mu of pastureland over the period. This land was only fertile for a few years and suffered from lack of rainfall. Already some 20 million mu of it has turned into desert. There are many other examples of the deterioration of pastures.[6]

Summary

We established in the preceding section that, as a result of the climatic extremes which they experience, the grasslands of China are an extremely fragile environment. The argument of this section is that there has been a steady deterioration of those grasslands. The deterioration arose from a number of forces, including unchecked cultivation, deforestation and overgrazing. As we shall discuss below, the problem of overgrazing has been exacerbated during the recent reforms because of a failure to create mechanisms for adequate control of access to pastures. However, the evidence cited above shows that the deterioration of the grasslands in China has also been underway for a much longer period. This reflects the tension between the claims of settled agriculture and the interests of the traditional nomadic people, a tension which is expressed in the encroachment of agriculture on the pastoral lands. Since 1949 that tension has continued under the aegis of the planning system, through the promotion of cultivation, the sanctioning of deforestation and the pressures for overstocking.

EFFECTS OF GENETIC CHANGES

Sheep-raising in northern China and central Asia has long been associated with the way of life of the nomadic pastoralists who now form part of China's national minorities. They developed coarsewool breeds suited to the environment which they raised for both wool and meat. Much of the wool was traditionally used for felt and carpets, and many of the minorities have developed a strong preference for the flavour of meat from local breeds.[7] While estimates of the size of flocks before 1949 must be treated with caution, the evidence suggests that, after the disruption of war, the number had decreased by as much as one-third compared to 1937 (see Table 8.1).

Since 1949 major efforts have been made to improve and change the genetic structure of China's flocks. The main trend has been a large increase in finewool and semi-finewool breeds. In 1954 the first domestic breed of finewool sheep —the Xinjiang Finewool Sheep, used for both wool and meat —was formally established after a breeding program begun in the 1930s, involving local coarsewool and imported Russian finewool sheep. The quality of wool was greatly improved. After systematic selection over 30 years, another breed, the Northeast Finewool Sheep, was established in 1967. In addition, a number of new finewool and semi-finewool varieties have been or are being bred in other regions. Over the long term, therefore, the dominance of coarsewool sheep has declined. Finewool, semi-finewool and improved sheep accounted for 41 per cent of total sheep numbers in 1987, compared to about 30 per cent a decade earlier and 14 per cent in 1970.

Table 8.2 shows the volume of wool production in China since 1973. Output increased rapidly after 1978 and yields in the 1980s were much higher than in the 1970s (1.43 kg per head average in 1973 and over 2 kg per head in 1987). This rise is partly explained by the higher proportion of fine and semi-finewool sheep.

Table 8.1 Number of sheep, 1949-88

	Total number (m)	Proportion of fine wool and semi-fine wool sheep (per cent)
(1937)	39.7	
1949	26.2	
1950	28.5	
1951	31.9	
1952	36.9	
1953	42.8	
1954	48.2	
1955	50.2	
1956	53.1	
1957	53.4	1.32
1958	50.4	
1959	61.9	
1960	61.4	
1961	60.8	
1962	64.1	
1963	69.7	5.9
1964	74.5	
1965	78.3	8.1
1966	na	
1967	na	
1968	na	
1969	na	
1970	85.6	13.9
1971	87.3	16.1
1972	88.0	18.7
1973	93.2	22.3
1974	94.7	23.6
1975	95.3	25.3
1976	92.7	25.4
1977	93.5	28.7
1978	96.4	30.4
1979	102.6	34.8
1980	106.6	36.6
1981	109.5	38.0
1982	106.6	38.3
1983	98.9	37.0
1984	95.2	35.4
1985	94.2	35.9
1986	99.0	37.1
1987	102.7	40.9
1988	110.6	na

Note: na — not available.
Sources: *China Agriculture Yearbook*; *China Statistical Yearbook*, various issues.

Finewool sheep average 3.5 kg per head (greasy) and semi-finewool sheep average 2.7 kg per head (compared to average yields of over 5 kg per head in Australia and New Zealand). Fine and semi-finewool accounts for about two-thirds of the clip (Table 8.2). Map 8.6 shows the regional distribution of wool output in 1987; Map 8.7 shows the regional variation in yields per sheep.

While there has been a significant change in the composition of the Chinese sheep flock towards finer wool production, there are significant issues to be resolved in the next stages of the breeding program. Some of these are reviewed in this section of the chapter. First, however, we shall briefly review the characteristics of the main types of sheep in China.

Coarsewool sheep

Traditional Chinese sheep breeds can be divided into three main types: Mongolian sheep, Kazakh sheep and Tibetan sheep (Zheng Peiliu, 1985). Mongolian sheep are fat-tail sheep, chiefly located in the northern Inner Mongolian prairie and in the western desert region of Ningxia and Gansu. Kazakh sheep have fatty buttocks and are distributed over the west Xinjiang desert regions and mountain areas. Tibetan sheep, which are a small-tail variety, are dispersed over the southwest Qinghai–Tibet Plateau at an elevation of 3,000 metres or more.

All three coarsewool breeds share several characteristics. First, they are adapted to the poor natural environment and to migratory or nomadic herding.

Table 8.2 Output of wool, 1973–89

Year	Total	Fine wool (mkg)	Semi-fine wool (mkg)	Share of fine & semi-fine wool in total (per cent)
1973	134.2			
1974	133.9			
1975	150.5			
1976	128.1			
1977	136.3			
1978	137.7			
1979	153.2			
1980	175.7			
1981	189.1	74.7	39.5	60.4
1982	201.8	88.1	41.7	64.3
1983	194.1	89.0	38.0	65.4
1984	182.8	85.6	34.4	65.8
1985	178.0	85.9	32.1	66.3
1986	185.2	89.8	32.1	65.8
1987	208.9	100.1	37.0	65.6
1988	221.7	110.7	na	na
1989	238.0	na	na	na

Note: na — not available.
Sources: *China Agricultural Yearbook*, various issues; *China Statistical Yearbook*, 1989.

Map 8.6 Distribution of sheep wool output by region, 1987 (per cent)

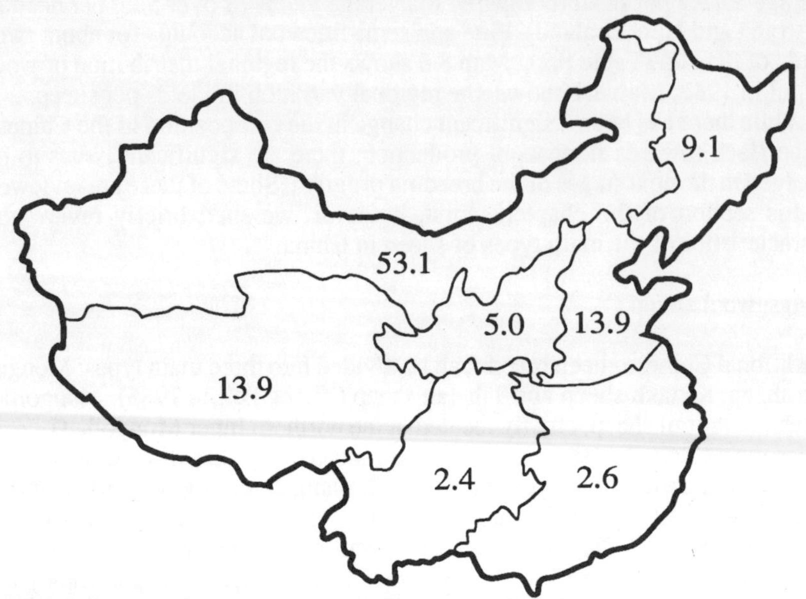

Source: *Agricultural Yearbook of China*, 1988.

Map 8.7 Average yield of sheep by region, 1987 (kilograms per sheep)

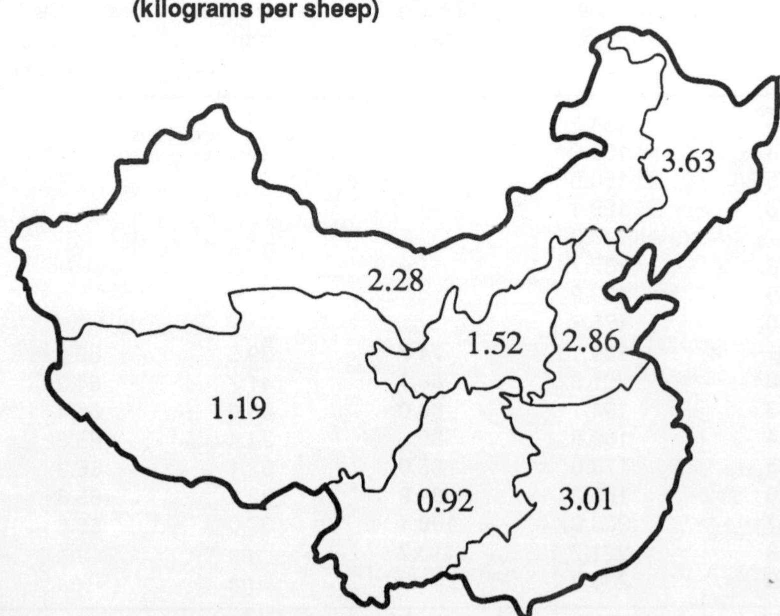

Source: State Statistical Bureau, China.

Second, they have a great ability to store fat. Third, the wool yield is low—only about 1 kg per head per year. As a result of variations in environment and long-term selection, these three basic types can also be further subdivided into a number of local breeds.

Because of the seasonal variations in plant growth in the pastoral area, the local breeds need to be able to store body fat to maintain their nutrient requirements. At the same time, in these climates, the local herdsmen also require meat with enough fat to protect them against the severe cold. These factors led to the development of the fat-tail sheep. In the farming area with a different environment, the character of the breeds is different from those grazed in the pastureland, and both fat-tail and short-tail varieties have developed. Han sheep, which are suitable for temperate and semi-humid areas, are an important breed in the agricultural areas in the Huang-Huai-Hai district. There is a fat-tail and a short-tail variety. Since the 1960s, however, these varieties have been crossed with improved breeds and the number of pure-bred Han sheep is declining. They have fine wool, a high reproductive capacity and fast growth. In the semi-pastoral and semi-agricultural zone, as discussed above, the main breeds, such as the Tan, are raised for fleece production.

Most of these three main coarsewool sheep types have been developed to produce finewool and semi-finewool by cross-breeding. However, some superior coarsewool types have been maintained as pure breeds to provide both meat and the coarse wool needed for carpet production.

Finewool and semi-finewool sheep

In the 1950s and 1960s a large number of finewool sheep varieties, mainly Merino from the Soviet Union, were introduced into China and crossed with local sheep to produce fine and semi-finewool. The use of imported breeds has always been important. As early as the 1930s the Caucas breed was introduced from the Soviet Union as the foundation for the Xinjiang Finewool sheep. Australian sheep were introduced from the mid-1960s. One of the first breeds to be introduced was the Polwarth. Now these sheep are raised mainly in the northeast and in Xinjiang where they are well suited to the local environment. According to trials at the Heilongjiang Animal Research Institute, a Polwarth ram crossed with a local Northeast Finewool sheep leads to significant improvements in staple length.[8]

The introduction of Australian Merino genetic material has also had some benefits (see Science Publishing House, 1977). In Xinjiang, for example, clean wool yields and fibre length have both risen. But, on the negative side, there has been a decline in the lambing rate and a decline in body weight by between 4 and 17 per cent depending on the extent of the cross-breeding.[9] The significance of these side effects is discussed further below.

Issues in the breeding program

Raising local sheep is much easier than raising introduced sheep, especially finewool and semi-finewool sheep, which require better feed and shed conditions. The latter require forage supplements and warm sheds in early spring for lambing. Finewool sheep produce more wool than coarsewool sheep and therefore need more feed. The wool of finewool lambs is shorter at birth than that of coarsewool lambs and their resistance to cold is lower. Shed facilities are needed to reduce the death rate. A consequence of this is that more labour is required to raise fine and semi-finewool sheep. Another important issue is the tension between wool and mutton production.

Mutton is an important joint product with wool in the Chinese sheep industry. As already discussed, most Chinese sheep are located in the north and northeast regions, where minority nationalities live in compact communities. Their preferred meat is mutton, the production of which is very important. The Mongolian, Kazakh, Tibetan and other nomadic minorities prefer the taste of mutton produced from local sheep breeds. Therefore, when improving sheep breeds, it is valuable from their point of view to maintain the meat characteristics of local varieties.

With the improvement of living standards, the demand for mutton is growing in other parts of China as well. In the period before 1978 pork consumption accounted for 94 per cent of total meat consumed, and mutton consumption amounted to only 2-3 per cent. After 1984 the pork share declined to about 84 per cent while the beef and chicken shares rose and the mutton share remained the same (China Agricultural Yearbook Editorial Committee, 1988). As tastes change and consumer preferences for variety in meat consumption increase, the demand for mutton will grow. Furthermore, mutton from older sheep and from those eliminated from breeding programs is becoming less acceptable on urban tables. Lamb meat is becoming more popular. The balance between wool and meat production is thus of considerable importance to the acceptance of new breeds. As noted above, the introduction of Australian Merino genes, for example, has had a negative affect on body weight and lambing rates. In this case, there is thus a trade-off between meat production and the quality of the wool clip.

The relatively higher costs of finer wool production and the impact of finewool sheep on meat yields suggests that the marginal costs of shifting to higher shares of fine and semi-finewool sheep in total sheep numbers are rising.

Other important issues in the breeding program are the age structure of the Chinese sheep flock, and the nature of flock management. The former is the basis from which the program has to develop, and the latter provides the framework through which it has to be realised. Average fertile ewes accounted for 51 per cent of total sheep numbers in China in 1986, and among fine and semi-finewool sheep, fertile ewes amounted to just 46 per cent (China Agricultural Yearbook Editorial Committee, 1988). This is much lower than the share of fertile ewes in more developed sheep-raising countries.[10] Too many old sheep and too few ewes will result in low productivity and high production costs. The flock population structure is in turn, however, related to flock management practices, and since 1978 those practices have been profoundly influenced by the introduction of the household responsibility system. The impact of that system on both pasture and flock management is therefore reviewed below.

THE HOUSEHOLD RESPONSIBILITY SYSTEM

The introduction of the responsibility system in these pastoral areas passed through several stages: the 'several fixed and one reward' system, contracting animals out to households, and finally selling the sheep to households.[11] The pattern of development was thus very similar to that of household contracting in the agricultural areas (see Watson, 1987). The 'several fixed and one reward' system involved assigning output, cost and quality targets to groups or individuals in return for an agreed share of collective income. Surplus output within the cost and other targets might either be kept by the producer or shared between the producer and the collective. Contracting animals to households meant that the households were given independent responsibility for the management of an assigned flock of sheep and agreed to deliver a fixed amount of product. In return, they received a share of collective income and had the right to keep any surplus output. As in crop-farming, however, the economic and technical pressures within this arrangement built up a momentum for ever greater household independence. In the case of animals, for example, there were many problems in defining the quality of the flock assigned to the household and the measures to be used to ensure that the quality was maintained. Furthermore, as young animals were bred, the question of ownership rights over the animals became even more unclear. Eventually, from 1983 onwards the flocks were either sold to households or transferred freely through allocations of equal value to each household (China, State Council Rural Development Research Centre, 1987).

Pastures were contracted to households according to sheep numbers, and the right to use the land was fixed for a long period—30 years for developed pasture and over 50 years for newly opened land in Inner Mongolia in 1983, and for over 30 years in Xinjiang in 1984 (China Agricultural Yearbook Editorial Committee, 1986: 338). Nevertheless, ownership of the land was retained by the collective or by the state, and the herders merely gained the right to use the land (though that right could be transferred to others and charges made for any investment in the land in the interim). By the end of 1982 sheep kept by households accounted for 61.7 per cent of the national total, a dramatic increase over the figures of 54 per cent in 1981, 46 per cent in 1980 and 33 per cent in 1978. By 1986, 88 per cent of all sheep were run by households (Chey, 1987).

By sweeping aside the collective system which had dominated sheep-farming for some twenty years, these changes had profound social and economic implications. In terms of production, the impact had both positive and negative dimensions.[12] On the positive side, the herders now had a strong incentive to look after their sheep. As a result, for example, numbers which perished through the effects of snow and cold weather fell dramatically. It is reasonable to assume that the efficiency of management increased, both as a result of incentive effects within the herding operation and through a process of self-selection whereby people who had a comparative advantage in wool-growing expanded their activities. Factors such as these may well have contributed to the large rise in sheep numbers from 1978 to 1981 (see Table 8.1).

It is significant to note, however, that once the households gained full ownership of flocks after 1982 and were no longer constrained by contracts to maintain flock numbers and specified levels of output, the total numbers of sheep declined.[13] As discussed in the next chapter, this full transfer of ownership of flocks is likely to have made the opportunity costs facing the peasants through the operation of the price and marketing system much more immediate. The result was that the peasants quickly reduced the size of their flocks even though the new flock management system gave them the incentive to be more efficient and productive.

Alongside the positive effects of the transfer to household management and ownership, however, there were also a number of problems. Sheep were commonly not allocated by skill levels or by a household's capacity to raise them but instead were allocated equally between each household in a collective on a per capita basis. This subdivision into small flocks also led to mixing of sheep types within flocks, such that finewool and coarsewool breeds ran together. In some instances, it also led to a mixing of the types of animals raised by each household. This has resulted in some decline in levels of specialisation and problems in the control of breeding. Because of inter-breeding between different sheep types, wool quality (both fineness of the fibre and being true to type) fell and there was an increase in the volume of coloured wool. It is estimated that as a result of this policy, the number of finewool sheep in Inner Mongolia fell by 1 million or 43.3 per cent between 1979 and 1984 (Liu Delun and Li Zhengqiang, 1987).

Economies of scale in herding were also lost by the new system. It is generally estimated that each herder can manage 100-150 sheep. The division of collective animals among households, however, meant that most flocks were under 100. A survey in late 1984 and early 1985 found that the total number of flocks had increased by some 60 per cent, with the largest flock being around 200 sheep and small flocks having 40 to 50. This had resulted in an increase of 50 per cent in the amount of labour needed to herd them.[14] Since labour is the largest cost in sheep-rearing, this meant a substantial increase in the total production costs for the industry as a whole. Although in some areas there have been attempts to combat this problem by encouraging the development of households specialising in sheep-farming, with larger flocks and larger allocations of pasture land, this process has not gone very far since in the relatively poor pastoral areas there is little alternative employment for the families thereby displaced (China, State Council Rural Development Research Centre, 1987: 50). Achievement of these scale economies is now a key issue facing wool production.

A related problem, of special significance for breeding and flock quality, was the provision of technical services to herders. The original veterinary and extension system was structured to handle a relatively small number of collective and state flocks. After the production responsibility system was implemented and the sheep flocks were dispersed, technical advisers found themselves faced with large numbers of small and scattered flocks. With limited facilities and a lack of funds and manpower, it was difficult to maintain services. Mobile services were limited by transport costs, and fixed position services could not satisfy the needs of dispersed flocks. Moreover, individual herders now had to bear the costs involved. They either could not afford to pay or were keen to reduce their costs. The original organisation was unable to cope and new organisations were not yet formed. Inevitably, breeding and artificial insemination programs were hampered. Herdsmen began to rely on their own rams, and controls over the genetic structure of the flocks began to decline.

INVESTMENT IN INFRASTRUCTURE IN PASTORAL AREAS

Over the last four decades the application and spread of new techniques has been handicapped by the lack of investment in infrastructure in the pastoral areas. As a result, pasture development has progressed slowly.

Grassland improvement

Despite the importance of animal husbandry in China's economy, government investment in it has been very limited. Agricultural investment as a whole has declined in recent years, falling from about 7 per cent of total state investment in all sectors to about 3 per cent (Watson, 1989). Investment in animal husbandry has been even lower, amounting to only about two-thousandths of the capital construction investment of the national economy; from 1949 to 1987 animal husbandry investment was 8.74 billion, which is only 1.6 per cent of the total output value of animal husbandry (542.15 billion) and amounts to 3.4 per cent of agricultural investment (255.04 billion) over the same period, an average of 2 fen (cents) for each mu of grassland (Wu Jinghua, 1987).

Some progress has been made in the development of grassland. Up to 1985, 100 million mu of improved grassland and 70 million mu of fenced grassland were established (Ai Yunhang, 1987). At the same time, varieties of grass-seed suited to planting in both the south and north of China were introduced and some experimental farms for planting grass and raising livestock were set up. These have played an important role in improving grasslands but the development in different areas has been uneven. Generally speaking, the pastoral areas, accounting for 40 per cent of China's territory, have not been an important focus for national investment plans. Conditions have changed little over this period and, as discussed above, seriously degraded pastureland exists in all areas.

According to one estimate, over the last four decades the gross value of animal products that Inner Mongolia provided to the state was 13.8 billion, but investment in animal husbandry, including operating expenses, was only 1.17 billion. The input was only one-twelfth of the output (Qian Fenyong, 1987). At the same time, investment in basic construction for the livestock industry was only about 180 million, an average of 13 fen per mu of pastureland. Qinghai Province is one of the poorest provinces, living on state subsidies and with little local revenue. During the period 1950 to 1985 investment in animal husbandry was only 125 million, averaging 25 fen per mu. Some 75 per cent of sheep herding households still make a living by migrating from place to place in search of water and grass (Ai Yunhang, 1987). Annual revenue is low in Xinjiang and many people depend on subsidies from the central government. From 1950 to 1985 investment in pastures was less than 1 per cent of the region's agricultural investment. The total investment for this period was only 224 million, an average of 19 fen per mu (Mohamed Yusupu, 1987).

The use of fencing to conserve and develop pastures also faces many constraints. The need for sheep to migrate between seasonal pastures, the fact that pasture is owned collectively by villages and only contracted to households, and the high costs involved in fencing the large number of household pasture areas all complicate the introduction of fencing at present. As stated at the outset of this chapter, there is conflict between the collective ownership of land and the private use of that asset by households. Households have strong incentives to maximise their use of the land and to minimise their investment in it. In most of the unfenced regions, the grassland is distributed in name to each household (according to the number of people in the household or the number of animals contracted), but in reality it is shared by the whole village. The sheep are moved to wherever the grass is best and are not confined to the grassland contracted to their owner. Whoever has the most sheep gets the greatest benefit from the pastureland used in common. By its nature, this form of grazing encourages misuse of the collective asset and exaggerates the degradation of pastures.

Veterinary resources and epidemic prevention

The shortage of funds has also restricted the improvement of livestock, the use of sheep-dipping, and the prevention and cure of diseases. There are two main reasons. First, there is a lack of funding for the basic technical sectors such as veterinary and epidemic prevention stations. Second, the herders cannot afford the expenses of sheep-dipping and epidemic prevention. In a pastoral area, characterised by continuous contact with other flocks and a mixed and extended grazing pattern, the immediate benefits to any one sheep herder of dipping are low, so there is no incentive to invest. A further aspect is that the veterinary station's task is to carry out a policy of 'prevention first'. Many herdsmen prefer to leave things to chance rather than pay for the service, and the administration cannot force them to take protective measures.

In most of the pastoral region education is very poor and there are few trained specialists. In Inner Mongolia each veterinary surgeon is responsible for more then 110,000 head of stock, and there is just one grassland specialist to 24 million mu of prairie; in Tibet there is 1.2 billion mu of grassland and 20 million head of livestock but only 17 grassland specialists, 22 animal scientists and 721 veterinarians (Ai Yunhang, 1987).

CONCLUSION

The quality of the grasslands and the maintenance of the genetic characteristics of the flock are critical to the growth in wool output in China. Both assets have seriously depreciated. The introduction of the household responsibility system without complementary initiatives on the management of assets of these types has in many ways intensified the problems.

The grasslands of China were already degraded by unchecked cultivation, overgrazing and deforestation. This process has occurred over the whole period since 1949. The issue of the quality of the grasslands is thus an aspect of a long term trend in China's animal husbandry. The problems were exacerbated in the 1980s when sheep flocks were distributed to households. Access to areas of grassland was allocated to households but there was no mechanism for enforcing property rights and ensuring that households maintained the quality of the asset. As a result, the grasslands were treated as a free input and, once the 'wool war' got underway, overgrazing accelerated.

The reforms also complicated the execution of new breeding programs, and even the maintenance of the quality of the existing flock. Larger flocks were fragmented at the time of the reforms and, in the smaller flocks, sheep varieties were mixed together. As a result, the quality of breeds and of the wool clip has fallen.

Over the last couple of decades the importance of finewool and semi-finewool sheep in total sheep numbers has risen. They now account for about 40 per cent of total numbers and about 65 per cent of wool output. However, the marginal cost of an increase in these shares seems high. This is because of the relatively higher costs of caring for finewool sheep in the harsh pastoral environment. Even more important are the opportunity costs in terms of meat production foregone. Both issues will become increasingly important to the herdsmen, mainly minority groups, and a point of conflict between them and the central planners, who seek to increase wool output. These are more recent examples of the long-run tensions between the nomadic communities and the interests of the settled provinces of China.

The basic constraints of environment and breeds are now also profoundly affected by the economic pressures of the systems of ownership and management. Incentives at all levels—government, village and household—are the key to the resolution of these issues, including more rapid institutional change to correct some of the property rights issues that remained unresolved.

We can review the options for resolving this issue by considering just two inputs into wool production; that is, pasture land and the sheep flock. In this discussion we will focus on management responsibility, and distinguish management from ownership, on the grounds that various types of management structures can, through the use of the appropriate contractual arrangements, be made consistent with various allocations of ownership rights.

Both the pasture land and the sheep can be managed either collectively or privately. Before 1983 both were managed collectively. After 1983 the management of the sheep was shifted to the household level; responsibility for the management of land was in principle also allocated to the household level but in practice the land was treated as a common input. As argued above, this situation makes the analytical apparatus of the 'tragedy of the commons' appropriate for this period. In these circumstances, we argue, the deterioration of the pastures, which has been a long-run feature of animal husbandry development in China, was exaggerated.

The question then is what are the appropriate institutional arrangements for the management of the sheep and pastures? Private management of the flock is less of an issue than the choice of a system for managing the pastures, since further reforms to marketing systems could provide stronger incentives to herder households to avoid the deterioration of the quality of their wool clip.

The management of land is more complex. For example, enforcement of private ownership by fencing would incur high costs and may be inappropriate where sheep follow migratory patterns. If that option is uneconomic, and given that the current arrangements leading to common access are also inefficient, then the solution may be to re-contract the management of sheep by, for example, combining them into larger flock sizes associated with defined territories in the pasture lands. A complementary action might be to include a fee for pasture use in the contract system. Recent reports indicate that this is already done in some parts of Inner Mongolia and is spreading to other regions. The system involves the classification of pastures according to grass output and recording the number of livestock grazed (Summary of World Broadcasts, Part 3, *The Far East*, FE/0867/B2/1, 12 September 1990). Nevertheless, such a system has high management and policing costs and still requires some mechanism for the reinvestment into the pastures of the revenue raised from the fees charged.

Our purpose is not to prescribe an ideal solution but to comment on the problems created by the current set of institutional arrangements. Furthermore, even if it were possible to identify a reasonably efficient allocation of responsibilities, it is not clear that the local governments have the fiscal incentive to take any action. The next chapter in this volume will thus turn to a more detailed assessment of economic incentives through an analysis of recent developments in wool pricing and marketing in China.

9 The 'wool war' in China

ANDREW WATSON AND CHRISTOPHER FINDLAY

INTRODUCTION

It was argued in Chapter 8 that the introduction of the responsibility system had some positive incentive effects in the sheep industry in China but also brought new problems for flock management, the provision of services and use of pastures. The fragmentation involved also raised costs so that the net effect of the changes was mixed. In other words, given the nature of wool-growing technology, the relationship between flock and pasture management and the degree to which management decisions by one household affect its neighbours, the institutional innovations resulting from the production responsibility system have not been the appropriate ones for the wool-growing sector. At the very least, the changes highlight the need for further significant reforms in the provision of services, the management of land, and the organisation of scientific and technological inputs. The problem is one of defining the appropriate location of responsibility for animals and pastures, which has implications for scale of operations and producer incentives.

This pattern of mixed experience in sheep-raising is an example of a fundamental issue underlying all of China's agricultural reforms, which, in institutional terms, can be regarded as movement along a continuum between centralised control and atomistic management by each household. Either pole position can have disadvantageous side effects and the optimum position may vary for different rural activities.[1] The changes reviewed in the previous chapter were those affecting the supply side of wool production through reorganisation of the institutions for production. An equally important issue has been the reform of pricing and marketing which shapes the economic environment for producers. If wool and meat prices and marketing arrangements had remained constant, the net effect of the institutional reforms might have been different than has, in fact, been the case. Since price and marketing changes were causing the relative profitability of different animal products to change, however, the end result of the institutional changes has also been modified. In turning to the second aspect of the reforms, therefore, it is first necessary to review the pricing and marketing of raw wool up to 1978.

WOOL PRICING AND MARKETING BEFORE 1978

The Ministry of Commerce was founded in 1952 with the aim of unifying the administration of domestic and foreign trade.[2] It was assigned authority over state commercial departments, and cooperative and private commercial firms, and was also involved in the administration of the prices of major commodities. In the early 1950s the Ministry established fifteen specialised companies, with central and local branches, to manage the purchase and sale of various products. Wool and other pastoral products were assigned to the Native Products Corporation. Over the following years, private commerce was merged with the state departments, which thereafter dominated all commercial activity. In the same period, the various supply and marketing cooperatives spread across the countryside were brought together into a related network under state control. The State Council established a central organisation and many intermediate level cooperatives were set up to manage the work of the cooperatives in their region. Although at various times thereafter the cooperatives were organised either as an independent network or as a sub-unit within the Ministry of Commerce, they mainly functioned as agents of the state commercial departments. Apart from their commodity coverage, they differed little from the rest of the state system. The quantity of raw wool purchased by the cooperatives in 1952, for example, accounted for some 24.2 per cent of total state purchases. Eventually, the supply and marketing cooperatives took over the main responsibility for the purchase of raw wool, which it then passed on to the appropriate state commercial agency.

At the same time as this centralised commercial network was taking shape, measures were introduced to control the distribution of agricultural products according to different categories. The first category consisted of vital commodities of major economic significance, commodities produced in particular areas but consumed widely and major export commodities. These were all subject to strict production, sales and distribution quotas. The second category consisted of commodities required for special purposes, other commodities produced in specific areas but distributed widely, and some commodities which were produced over wide areas but which had to be supplied in a guaranteed way to particular places. This category was also subject to strict controls, especially in the areas of specialised production. The third category consisted of all remaining commodities. Although there were subsequent modifications in the list of commodities in each category, in the strictness of the controls exercised, and in the commercial networks charged with handling the distribution of each commodity, this basic framework of commodity classification has remained in place until the present. Raw wool has consistently been listed in the second category and subject to firm state controls over purchasing, price and distribution. Until the reforms began, therefore, raw wool marketing was monopolised by the supply and marketing cooperatives (with the Native Products or Pastoral Products Corporations acting as agents in some areas) and strictly regulated by the state.

The centralisation of commercial networks and the introduction of planned controls over commodity management was accompanied by increasing state regulation of pricing. In 1955 the Fifth National Conference on Prices decided that the responsibility for the Ministry of Commerce was: to control general price levels and to administer prices in the domestic market, to determine the difference between purchase and sale prices, to determine the price variations for different regions, seasons, qualities, and to determine relative prices between different products; to stipulate and adjust the basic purchase, wholesale, intermediate and retail prices of major commodities and the domestic prices of imported products; to stipulate the pricing methods for the transfer of goods within the commercial system; to study the principles of pricing used in commercial and other relevant systems; and to draw up the regulations for price management in the state system (Commercial Economics Research Institute, Ministry of Commerce, 1984: 33–4).

Thereafter, although there were changes in the agencies which administered the planned price system, the basic principles for price management remained the same. Price changes and adjustments in relative prices were announced through the State Council and reflected the priorities of the administrative and planning systems rather than supply and demand and changes in productivity and costs. At various times, the state did attempt to use prices as an incentive to producers, with higher surplus prices for sales above the obligatory state sales quotas and occasional experiments allowing producers to sell surplus output of commodities in the strictly controlled categories at negotiated prices, either to the agencies charged with managing the commodity concerned or to other departments altogether. The overall trend, however, was to reduce regional, seasonal and quality price differentials and to establish a single basic price at which the bulk of production was purchased by the state. During the Cultural Revolution from 1966 to 1976, for example, prices were frozen for long periods.

As a category two commodity, raw wool prices were among those tightly controlled. The price levels were calculated on the basis of an experimental price which was originally related to the market price for wool in the period shortly after 1949. Wool prices for selected years since 1952 are shown in Table 9.1. Between 1953 and 1965 the average sheep wool price fell from 2.38 yuan per kg to 2.18 yuan. From 1966 to 1972 it remained fixed at 3 yuan, and by 1977 it had risen to 3.28 yuan (China, State Statistical Bureau, 1983: 482). From 1952 to 1977 the wool price (mixed average price for state purchases) increased at an annual average rate of under 2 per cent a year. Nevertheless, since other agricultural product prices did not change much either, the relative price level for wool remained much the same.

The final point to note in this overview of wool marketing trends before 1978 is that wool producers did retain a proportion of wool output for their own use. That wool was used primarily for making felt or carpets for their homes. Unlike cotton, which Chinese peasants used to make homespun garments and of which the volumes retained by producers were regulated as part of the quota system, however, the amounts of wool retained by producers were not considered large enough to warrant specific controls. Moreover, most of that wool was coarse rather than fine and part of the demand for clothing was covered by the use of whole skins to withstand the rigours of the harsh winters. Nevertheless, the keeping qualities of wool meant that households could maintain stocks for their own use or while waiting for a change in demand. In Xinjiang, for example, over

Table 9.1 Sheep and wool purchase prices, China, 1952–86

	Sheep wool purchase price (yuan/kg)	Sheep skin price (yuan/piece)	Sheep meat price (yuan/head)	Grain price (yuan/ton)
1952	2.08	3.5	8.8	138.4
1972	3.00	4.0	13.3	256.0
1975	3.06	3.7	14.0	254.4
1978	3.40	4.2	14.7	263.4
1979	3.40	4.5	17.7	330.7
1980	3.43	4.5	21.7	360.6
1981	3.48	4.7	25.6	381.7
1982	3.58	4.7	25.0	392.2
1983	3.66	4.7	27.8	392.6
1984	3.73	5.3	29.0	395.1
1985	5.04	9.4	39.1	416.4
1986	6.01	16.6		465.9

Sources: 1952–1983, *Zhongguo Maoyi Wujia Tongji Ziliao 1952–1983* (Zhongguo Tongji Chubanshe, 1984: 451, 452, 455); 1984–1986, *Zhongguo Nongcun Tongji Nianjian 1987* (Zhongguo Tongji Chubanshe, Beijing, 1987: 144).

the years 1977 to 1981 it is estimated that wool purchases averaged 88.62 per cent of output and skins averaged 61.84 per cent (Li Yuxiang, 1986: 214).

WOOL PRICING AND MARKETING REFORMS AFTER 1978

In the price reforms introduced after 1978, it was stipulated that surplus output of agricultural commodities in the first and second categories could be purchased at negotiated prices, except for critical industrial raw materials, pastoral products, major vegetables and some Chinese medicines (Commercial Economics Research Institute, 1983: 369). According to these rules, therefore, wool producers could only get the basic purchase price imposed by the state and there was neither an over-quota price nor a negotiated price for raw wool. The supply and marketing cooperative remained the sole commercial agency for wool purchasing, and, as before, the state price depended on the administrative judgement of the commercial and price departments.

After 1978 agricultural prices were also increased as part of the overall program to speed up agricultural development.[3] This is evident in Table 9.1, which shows that the grain price rose by 26 per cent in 1979, by another 9 per cent the following year and by 6 per cent in 1981. Wool prices were slow to move, however, increasing by 3 per cent in 1978, around 1 per cent in 1980 and 1981, and by 2 per cent in 1982. In effect, they hardly changed at all in nominal terms, and relative to the grain price, which is used here as an indicator of prices for other agricultural activities, the price of wool fell until 1982. This restriction on the raw wool purchase price at a time when urban incomes were rising and the demand for woollen clothing was increasing provided a further strong incentive for the expansion of wool spinning and weaving capacity. Raw material prices remained low and the profitability of the processing industry was high. It is not surprising, therefore, that this attracted considerable investment in wool processing by local levels of government outside of the state plan.[4]

Contrary to the picture for wool, however, mutton prices rose by 20 per cent in 1979, 23 per cent in 1980, and then by another 18 per cent in 1981, after which they levelled off for several years. Mutton prices also rose faster than grain prices over the whole period, except for 1982. Thus the profitability of using sheep for meat increased rapidly. This may partly explain the rising stock numbers up to 1981 (see Table 8.1). To some extent, therefore, the growth in wool output may have been a by-product of the growth in sheep numbers to produce meat. Of course, the stockholding decision depends on returns to both wool and meat, but the major changes up to this time were the increases in the meat price. In Inner Mongolia before 1979 the wool/mutton price ratio stood at 1:3, a relative price which favoured wool production. After the 1979 adjustment, the ratio was reduced to 1:2.3, a shift which might be expected to work against wool production from an existing sheep flock. When meat prices were completely decontrolled in the autumn of 1984, mutton prices rose dramatically and the ratio was reduced to almost 1:1. Consequently, mutton output rose quickly and wool output declined. Between 1980 and 1985 mutton output in the region grew by 33 per cent while wool output grew by only 1.25 per cent (Wang Bingxiu, 1987: 13).

This discussion of the marked changes in relative pricing over the period 1979–84 raises the question of why the relative price for wool was allowed to decline so abruptly. The simplest explanation is a combination of oversight and a failure to anticipate the changes in producer behaviour that would result from the introduction of market exchange and from a decentralisation of production management which made producers sensitive to opportunity costs. The strong fluctuations in the output of various agricultural products during the 1980s and the decline in grain production after the 1985 price adjustments[5] all demonstrate the difficulties faced by the Chinese government in coming to terms with managing the new pricing system. Nevertheless, one of the major constraints in the price reforms to date has been the concern to protect urban consumers and urban industries from rapid price increases. Maintaining controls on wool prices, therefore, might have also been part of a conscious effort to sustain the profitability of the state-owned industries in the coastal regions. Seen in that light, the lack of concern for the relative price of wool might also be understood as a reflection of the long-run relationship between agricultural and pastoral interests discussed in the preceding chapter.

In this situation, a number of factors may have contributed to the decrease in the size of the sheep flock after 1981 (see Table 8.1) (assuming that problems with the statistical reporting system because of the introduction of family farming cannot account for the total decline). One explanation is that producers found the meat price so high that they increased the kill rate to a level whereby the flock size actually fell. Alternatively, it could be that the slowdown in the rate of growth of the meat price between 1981 and 1984 led them to believe further rises were as likely and thus it was not as worthwhile holding onto the sheep. In addition, as discussed in the previous chapter, over this period the ownership of sheep was passing to individual households. As a result, they were becoming more sensitive to the opportunity costs of different activities. Furthermore, it was not only the wool and meat prices which mattered but also the returns on time and effort in other rural activities, including off-farm work. The decline in sheep numbers, therefore, could have reflected their assessment of the price and cost structure they faced and the new freedom they had to respond to that assessment. In other words, the price ratios appeared to have had a very significant influence on wool production and wool sales.[6]

One further important factor here is that there are commonly several prices operating for each commodity. These include the state base price, the surplus price, the negotiated price and the free-market price. As a result, the marginal price facing the producer depends on the choice of which market and at which price level the next increment of output is traded. Furthermore, the relative price ratio between competing products may be different at each price level. Since mutton has been subject to less controls than wool, the flexibility on the meat production side is greater. Prices in Table 9.1 are for official purchases in the state system. The grain price, for example, is the official price, not the negotiated price. There were other prices for meat and grain, much higher than those in the table. In the initial period after 1979 there was no over-quota or negotiated price for wool, and it was not until 1983–84 that effective over-quota prices for wool began to emerge. In some areas, textile enterprises began to form direct links with producers and offered prices slightly higher than the state price (Gao Xiaoming, 1984: 54–5). Furthermore, private traders and other enterprises were also gradually beginning to enter the market. An example of private trading in 1987 illustrates how the process took place. A group of villagers in Yulin Prefecture, close to the Inner Mongolian border, decided to trade in wool. They hired a lorry and travelled across the border, where they purchased wool from producers at 12–14 yuan per kg. Returning to Yulin, they sold it to the local supply and marketing cooperative at 18–20 yuan per kg.[7] Until the price and marketing system was liberalised in 1985, such trading was more or less illegal. Nevertheless, it is easy to see how small-scale traders might aggregate purchases from individual producers and sell their wool to processors or marketing cooperatives for a profit. The opportunities for middleman profits escalated the extent of the wool war.

The immediate response to these changes, however, was not a positive one for wool production. As noted above, the relaxation of meat price controls in late 1984 led to a rapid increase in meat prices so the relative price levels were moving against wool in any case. The peasants may also have anticipated that the relative price of wool could not keep falling, and for the moment preferred to store their wool rather than sell it. Furthermore, by its nature, sheep-raising requires time to respond to changes in the economic environment; flocks cannot be built back up immediately.

An additional issue was that, despite the growing demand for raw wool, the state was having difficulty in storing it. As a result, the commercial departments were not keen to buy. According to one report, in October 1984 state stocks of domestic wool stood at 50,000 tonnes (probably a very approximate estimate). The commercial departments were burdened with carrying the cost of the stock and were anxious to sell it (Zhao Zekun, 1986). This large stock was the result of both the increase in wool purchases of the previous years and the increase in imports after 1978. The combination of peasant reluctance to produce and the commercial departments' reluctance to buy, meant that in 1984 the volume of raw wool purchased by the state system fell (see Table 9.2). Imports also fell that year by about 20 per cent.

This reduction in purchases and imports seems paradoxical in the context of

Table 9.2 Raw wool production, consumption and purchases, China, 1978–88

	Raw wool production mkg[a]	Total purchases mkg[b]	Raw wool imports mkg[c]	Raw wool availability mkg (imports and purchases)	Domestic purchases as a % of availability
1978	137.7	138.2	10.3	148.5	93
1979	153.2	149.1	16.9	166.0	90
1980	175.7	161.0	37.4	198.4	81
1981	189.1	170.1	59.1	229.2	74
1982	201.8	179.9	87.0	266.9	67
1983	194.1	177.7	88.4	266.1	67
1984	182.8	154.8	55.8	210.6	74
1985	178.0	219.0	113.4	332.4	66
1986	185.0	218.0	152.2	370.2	59
1987	208.9	205.0	152.5	357.5	57
1988	221.7	193.4	187.4	380.8	51
1989	238.0	na	104.4	na	na

Note: na — not available.

Sources: a 1978–79, *Zhongguo Nongye Nianjian 1980* (Nongye Chubanshe, Beijing, 1981: 118); 1980–85, *Zhongguo Noncun Tongji Nianjian 1986* (Zhongguo Tongji Chubanshe, Beijing, 1987: 94); 1987–88, *Zhongguo Tongji Nianjian 1989* (Zhongguo Tongji Chubanshe, 1989: 214); 1989, *Renmin Ribao*, 21 February 1990: 2.

b 1978-83, *Zhongguo Maoyi Wujia Tongji Ziliao 1952–83* (Zhongguo Tongji Chubanshe, Beijing, 1984: 121); 1984–86, *Zhongguo Nongcun Tongji Nianjian 1987* (Zhongguo Tongji Chubanshe, Beijing, 1987: 122); 1987–88, calculated as 96.7 per cent of all purchases (including goat wool) which was the average ratio for the years 1981–86 when the range was 96.1 per cent to 97.7 per cent.

c 1978–80, (Foreign Trade Department figures), *Zhongguo Tongji Nianjian 1984* (Zhongguo Tongji Chubanshe, Beijing, 1984:413); 1981–83 (Customs figures), *Zhongguo Maoyi Wujia Tongil Ziliao 1952–83* (Zhongguo Tongji Chubanshe, Beijing, 1984: 480–81); 1984 (Customs figures, *Zhongguo Tongji Nianjian 1986* (Zhongguo Tongji Chubanshe, Beijing, 1986: 572); 1985–86 (Customs figures), *Zhongguo Tongji Nianjian 1987* (Zhongguo Tongji Chubanshe, Beijing, 1987: 600); 1987, *China's Customs Statistics* No.1, 1988: 32; 1988, *China's Customs Statistics* No.1, 1989; 1989, *China's Customs Statistics* No.1, 1990.

the growing demand for wool. In effect, several factors were working in opposite ways at the same time. First, in 1984 the structure of prices did not favour wool. Even though the price was rising and the opportunities for sales outside the state price were growing, the market was still substantially under state control. Second, the Chinese economic system is one where producers, commercial departments and end-users are each subordinate to separate management systems, and each system faces its own range of priorities, costs and prices. The textile industry required wool but the commercial departments could not afford to buy and store it. The producers were not dealing directly with the users but with a range of purchasing agencies subject to administratively determined prices. There was thus a lack of direct linkages between producers and end-users. Third, the organisation of production was itself passing through the major changes associated with the introduction of the production responsibility system as discussed above. And fourth, the price and marketing system was under pressure to change. The old state controls were breaking down but there was no effective system to replace it. As 1985 began and the full impact of the extra-plan investment in spindles and wool textile production began to be felt, the stage was set for the wool war.

THE WOOL WAR

The decision which triggered the wool war and brought the interaction to a head of the various factors analysed above was the abolition of the unified purchasing system at the beginning of 1985.[8] The old centralised marketing and price system was replaced by purchase by contract or through the market, and this time the reform included wool. The result was an explosion of competition in the purchase of wool. Although it was intended that the state contracts should be filled first, before wool entered the free market, commercial departments, wool processing enterprises and private dealers all competed for supplies. Indeed, the ability to store wool probably introduced more dealers into the free market when it was actually set up than would have been the case for a more perishable product. The situation rapidly got out of control, with different enterprises and different levels of government offering various inducements and incentives. The result was that wool became subject to a range of prices, including the base price, negotiated prices, clean wool prices, weighted average prices, proportional surplus prices, and free-market prices (Liu Deyou, 1986a: 2). The state found it difficult to purchase wool at the base price and was forced to raise the raw wool price substantially in 1985 and again in 1986.[9] In 1985 the price rose by 35 per cent over the previous year. Purchases in that year far exceeded production (see Table 9.2) as peasants ran down the stocks of wool accumulated in anticipation of the marketing change. Presumably, part of the excess can also be explained by the fact that a proportion of output was no longer recorded in the production statistics.

As these price and market changes were taking effect, new policies for local processing were introduced by the governments of the main wool-producing areas (Gansu, Qinghai, Xinjiang and Inner Mongolia), which were anxious to expand their own wool-processing industries as a means of increasing their income. The policy became known as one of 'own production, own use and own sales'. In Qinghai the wool textile industry accounted for 20 per cent of provincial revenue, in Xinjiang for 10.41 per cent, and for over 8 per cent in Gansu and Inner Mongolia (Liu Deyou, 1986b: 2). At both provincial and county levels, new investment was taking place. During 1986, for example, Gansu was adding 16,384 spindles and Xinjiang was building 25 textile factories and some 15,000 spindles. These increases meant that most of the local wool production would be needed to satisfy local processing capacity (Liu Deyou, 1986b: 2). Inevitably, this led to increasing local reluctance to supply outsiders. Purchased wool was held in local stores and not sold to outsiders despite the storage costs involved (Liu Deyou, 1986a: 2). Xinjiang, for example, placed a levy of 30 yuan per tonne on 'exported' raw wool (Beijing Agricultural College Research Group, 1987: 292–302). This situation also applied to the relationship between the counties and the other levels of leadership within a province or region. A report from Inner Mongolia of early 1988 quotes the leaders of a banner (the equivalent of a county) as telling the authorities from an Inner Mongolian city that they were determined not to sell their cashmere wool and would hold on to it for local processing: 'Unless you sack us, you can't have it!' (Liu Deyou, 1988: 2). By 1987 this had led to the central authorities attempting to reduce the amount of wool used within the main producing areas from 53 per cent of raw wool supplies to 40 per cent, but, as the report from Inner Mongolia suggests, there are clearly many difficulties in the centre enforcing such regulations through provincial level authorities at a time when many powers are being decentralised. As a result of these processes, by the end of 1988 an estimated 2.26 million spindles were operating with an annual capacity of around 220,000 tonnes of clean wool against domestic output of the level of around 50,000 tonnes (clean) (He Long, 1990). Another estimate suggested that by early 1990 spindles reached 2.7 million (Men Xiuqi, 1990).

Indeed, this decentralisation of authority was one of the major engines driving the wool war, and it was also affecting all other parts of China outside the wool-producing areas. The key issues here were the introduction of financial contracting down to county levels after 1980 and the massive growth of investment outside the plan.[10] Under the new arrangements, each level of government makes a contract with the next level up to meet certain income and expenditure targets. If an administrative level is able to generate additional revenue outside the contract, this is shared with its next superior level at an agreed ratio. The surplus income retained at the lower level (which in some cases may be all such income) can be used to cover new investment or other expenditure at its own discretion. One of the major aims of this change is to make lower levels of government more efficient in budget management. The corollary, however, is that they also have a strong incentive to promote local investment outside the plan since this leads to rapid growth in their income. In other words, the financial contracting system means that the economic interests of each level of government are sharply identified, and there is strong competition to ensure that the returns to the local administration are maximised. Local governments in all areas began to invest in industries, which increased the value added to be gained from processing local raw materials, and at the same time tried to restrict the flow of those materials to other regions. They also invested heavily in the textile and other labour-intensive industries, even if they did not produce the raw materials themselves.[11] The result was the emergence of a succession of 'commodity wars' over such things as wool, silk and cotton, whereby local governments, production units and merchants competed for supplies. By 1988 this led one Chinese commentator to observe that 'the current market is one where the feudal lords contend, regions put up blockades, trades practise monopolies, and the wars are ceaseless' (*Zhongguo Shangye Bao* [China Commercial Paper], 23 August 1988).

The war thus had a marked effect on the behaviour of local administrations. In some cases it involved the setting of local production quotas, with fines or budget cuts for those who failed to meet their targets. Alternatively, steel, cement and timber supplies might not be issued, or cadres' wages might be withheld and the cadres themselves summoned for investigation (Ai Yunhang, 1988: 31). In other areas it took the form of organising the militia to blockade the transport which was taking the wool out of their district, and, according to some accounts, it was these actions which led to the coining of the term 'wool war' (Beijing Agricultural College Research Group, 1987: 292).

The intensification of this war over wool supplies was expressed in a number of ways. Inevitably, it had a strong effect on prices. Price rises in 1985 were reported to be generally over 50 per cent (Zhou Guohua et al., 1986: 32). By July 1986 the price of fine wool had risen by 70 per cent in Gansu, 19 per cent higher than the top limit stipulated by the provincial government, and the price of coarse native wool had risen by 112 per cent (Liu Deyou, 1986a: 2). By 1988 it was reported that Xinjiang wool prices were at about the same level as the import price and Inner Mongolian prices were even higher (*Zhongguo Fangzhi Bao* [China Textile Paper], 6 October 1988). Furthermore, price inflation was taking place at each level of trading so that local, county, prefectural and provincial supply and marketing cooperatives were all increasing their mark-ups, like the private merchants (Ai Yunhang, 1988: 32). Another feature was an upsurge in the adulteration of wool with dirt, oil, sugar water, and sand being added to increase the weight of the wool sold. This took place at all levels, beginning with the peasants and involving merchants and lower level purchasing cooperatives (Ai Yunhang, 1988: 31).[12] In one case in Gansu, some enterprising individuals set up 'processing workshops' to help the peasants adulterate the wool (Liu Deyou, 1986a: 2).

Alongside these efforts to 'increase' the amounts sold, there was also a decline in quality controls by the basic level units trying to purchase as much wool as they could, and an inflation of the amount of wool sold as fine quality. In Gansu some purchasing units bought all wool as a single grade. In Xinjiang in 1985 each variety of wool was classified one grade higher. In Qinghai first grade fine wool accounted for 40 per cent of purchases in 1983. By 1986 this had risen to 70 per cent (Liu Deyou, 1986a: 2). Furthermore, the commercial departments also used this as an opportunity to boost their income. In Xinjiang in 1985 only 4.4 per cent of raw wool purchases was classified as 'super-fine'.

When the wool was sold to users, however, 'super-fine' accounted for 17.4 per cent! (Beijing Agricultural College Research Group, 1987: 293–4).

The regional conflict for wool supplies helps explain why the wool war was maintained during the years 1985 to 1988 at a time when the raw wool price was rising and the rents obtained by processors from the difference between the low raw material price and the high end-product price were thereby declining. Several factors were working in concert to reduce the impact of this fall in rent income. To begin with, many of the investment decisions to move into wool-processing were taken in 1984–85, when profitability was high. These investments were entering production even as raw wool prices rose. Second, the conflict for value added between local governments and between interior and coastal provinces meant that, even if profit margins were declining, there was still competition for the raw material. Local governments could hope to use their administrative powers over local markets and producers to ensure a continued supply at favourable price levels. Finally, consumer demand remained buoyant until after the panic buying of 1988. Processors anticipated continued growth in demand and an increase in free-market prices for end-products which could absorb the increase in raw material prices. As indicated below, the squeeze on profitability came as much from the drop in clean yield from raw wool as from the rise in price of the latter. Ultimately, it was not until the credit squeeze of late 1988 and the ensuing drop in consumer demand that the pressures contributing to the wool war abated. The absence of a national market and the potential for administrative intervention thus counteracted the economic forces which might have been expected to flow from a drop in the rents obtained from processing raw wool.

The effect of the wool war on the wool textile industry was equally profound. The rise in prices and the decline in quality added to the problems created by the excess processing capacity. To begin with, the adulteration of wool added to transport and handling costs. At the same time, it led to a decline in the clean weight yield of scoured wool. By 1987 this had dropped to around 30 per cent (Ai Yunhang, 1987: 32) or even 20 per cent (Men Xiuqi, 1990). The lack of concern for quality and the mixing of wool grades also made it very difficult for processors to maintain a steady flow of production at the correct quality levels (Liu Deyou, 1986a: 2). As a result of these problems, there were calls to find ways of directly linking producers and processors, including the idea that processors might invest in pasture and sheep-rearing. Experiments of this kind were under way in Xinjiang for some years (Ai Yunhang, 1988: 32). In addition, in 1988 experimental wool auctions were held in Urumchi, Huhehot and Nanjing to enable processors to have a direct say in the amount they were prepared to pay for the quality of wool offered (*China Daily*, 28 August 1988: 1; *Renmin Ribao* [People's Daily], 29 September 1988: 1). It is likely that the experience of the wool war will continue to generate calls for reforms along these lines.

The increase in prices for raw materials also led to a reduction in industry profitability. By 1988 this was beginning to reach serious levels. Between 1985 and the first half of 1988 the profit rate of four major wool textile factories in Beijing had dropped from 19.16 per cent to only 0.19 per cent and the wool textile industry of the city as a whole was expected to make a loss (*Zhongguo Shangye*

Bao [China Commercial Paper], 6 October 1988). A similar prospect faced Tianjin, and Shandong was only avoiding losses because it still had stocks of 1987 wool. Furthermore, many of the most efficient and modern factories could not get supplies because of the competition from the large number of workshops established by local levels of government. These small factories enjoyed tax relief advantages, had lower costs, and, though producing poorer quality products, were able to offer lower product selling prices to compete with the larger enterprises. As a result, they could afford to pay higher prices for the raw wool (Liu Deyou, 1986a: 2). One of the net effects was that a higher proportion of the raw material was wasted because of the lower quality technology used in the local enterprises.[13] A similar chain of events occurred in the cashmere processing industry, and during 1988 cashmere prices went up by the month from 400 yuan per kg in March to 760 yuan in July (Liu Deyou, 1988). Ultimately, therefore, the wool war also led to a decline in the efficiency with which the raw wool was used.

Finally, as discussed above, the wool war also led to a rapid growth in wool imports. These were particularly significant for the coastal processing industry. Imported wool, though until recently more expensive than the domestic product, was supplied in reliable and standard qualities and thus still enabled a good profit to be made. In addition, the state commonly subsidised the price of the imported wool to the processors so that the price differential was less significant than it might have been. This subsidy amounted to some 5,000 yuan per ton in 1986 using the 1982 RMB/US$ exchange rate (Ai Yunhang, 1988). As might be expected, the growth in imports has led to discussion of whether the processors' preference for imported wool is acting as a disincentive for domestic wool producers. Also, the growing cost of imports meant that by 1988 a system of import quotas and import permits was introduced to restrict the flow. Importing provinces had to apply to the State Planning Commission for a quota and to the Ministry of Foreign Economic Relations and Trade for a permit before signing contracts.[14]

MARKETING REFORMS[15]

By 1988–89 the many problems that the conflict over wool was creating were leading to a new round of reforms. As noted in the previous section, initial experiments were under way to explore how to change the marketing system to one of auctions based on clean yield. Processors were being encouraged to develop direct links with producers, as a means of both reducing commercial costs in wool trading and enabling some of the profits from processing to flow back into wool production. The conflict, however, between state and local industries and the coastal provinces and the interior wool-producing regions for the raw material remained intense. It was reflected in the steps taken by areas such as Xinjiang and Inner Mongolia to control the outflow of wool by imposing 'export' taxes and by physically preventing it from being removed. By late 1988, therefore, the situation was still one of conflict over supplies, low quality of the domestic product, rising domestic wool prices and heavy reliance on imports.

Experiments with auctions and direct dealings between producers and processors were still in their early stages.

This overall situation was changed dramatically by the credit squeeze and deflationary policies adopted in 1989 in an attempt to control inflation and the general over-heating of the economy, and by the reassertion of planning priorities after the events of June. The squeeze on credit meant that both local and state-run enterprises had less working capital to buy raw wool. Both imports and domestic purchases dropped dramatically. At the same time, consumer demand fell as inflation declined, savings went up and overall income growth was slowed down. During 1989 sales of wool products declined, commercial departments reduced their stocks, and production enterprises found themselves with growing inventories of finished products. From the wool-producers' point of view, this led to a drop in demand for wool and a drop in prices. The state wool purchasing agencies also found that they were unable to sell their stocks of raw wool. In many areas these agencies did not pay for the wool but issued i.o.u.s to the herders.

Inevitably, this undermined producer confidence, and this situation was compounded by the drought conditions which hit many pastoral areas during the year. Pasture growth was insufficient to enable good storage of hay for the winter. Combined with the lower prices paid for wool in 1989, the end result was that many sheep-raisers were contemplating the slaughter of flocks prior to the 1989 winter (*China Daily,* 14 November 1989).

The impact of the 1989 credit squeeze on wool demand inevitably meant that the prices paid for wool came down from the highs of 1988. In that year, for example, the prices offered per kg of clean wool in the Urumchi auction in Xinjiang were around 60 yuan. By contrast, the top prices obtained for Xinjiang wool at the Beijing auction in September 1989 were around 43 yuan. At the same time, the prices paid to producers also dropped. According to the manager of the Tianshan Woollen Mill, after rising by 200–300 per cent between 1986 and 1988, prices stabilised and even dropped a little by the end of 1989. In Fukang County, Xinjiang, for example, the price for mixed grade greasy wool offered to herders in 1989 by the Animal Husbandry Bureau trading company was reported to be 10.50 yuan to 14.50 yuan. This compared with prices of 16–18 yuan in 1988. The 1989 price offered by the Tea and Animal Husbandry Products Company (the Xinjiang equivalent of the supply and marketing cooperative system) in Fukang for mixed grade wool was reported to be as low as 8 yuan. Furthermore, it was also said that this system had not paid in cash but in i.o.u.s to be redeemed later. While all these prices may not be entirely comparable (the grades of wool involved may not be the same), they are indicative of the drop in prices paid to the producer that had taken place.

A result of this change was that the wool/meat price ratio was beginning to move in favour of meat again. In addition, the downturn in the processing industry meant that much of the 1989 clip was being held in stores. It was reported that the commercial systems were refusing to purchase wool from herders in some parts of Xinjiang.

Reports on wool stocks and prices during early 1990 indicated that these trends had continued and were pointing towards a further downturn in raw wool production.[16] By February wool stocks in the main producing areas were said to be around 120,000 tonnes, with a further 40,000 tonnes in the hands of producers. Of the stock in the state system, some 20,000 tonnes were said to be 1988 wool.

Purchasing agencies were no longer able to buy and prices were dropping. Whereas wool with a clean yield of around 30 per cent had fetched 15 yuan per kg or more in 1988, the price of 40 per cent clean yield wool in 1989 had fallen below 12 yuan. By mid-1990 in Inner Mongolia the minimum price of 6.32 yuan per kg for top grade wool and 5.54 yuan per kg of improved wool could not be guaranteed. The supply and marketing cooperative could not store the wool and also lacked the capital to buy it. Nationally, the value of wool stocks was estimated to be over 1 billion yuan, with interest costs of 100 million yuan and losses to the supply and marketing system of 150 million. Not surprisingly, herders were beginning to slaughter sheep, with a 20 per cent drop in numbers in the three winter months in northern Shaanxi. In Inner Mongolia the low price for wool meant that it had become cheaper to use wool for padded quilts than to use cotton! Given that wool output was expected to increase further, it appeared likely that the price would fall further, that even more stocks would be left with producers and that producers would move out of wool production.

Wool pricing in China is still changing. The auctions held in Urumchi, Beijing and elsewhere have begun to shift the wool market towards pricing based on quality and clean yield and towards direct interaction between the textile industry and the wool producers. Over the long term, reforms along these lines will bring significant changes to wool pricing.

First, they will reduce middleman costs. By the time the wool passes up through the state marketing system from the purchasing agencies at the basic level, the handling costs result in a mark-up of around 30 per cent on the original price. The manager of the Tianshan factory, for example, reported that he preferred to buy his wool at county agencies since the mark up at that level was only 21 per cent compared with 23 per cent at the Xinjiang regional level. One observer also argued that one of the contributing factors in the high prices paid to producers in the 1988 auctions was the fact that the purchasers did not have to pay the commercial mark-up and, in the competition to buy wool, that component in wool prices was transferred to the producer.

A second gain from such reforms is that they will help improve grading and price differentials based on wool quality. This change is fundamental if there is to be any hope of encouraging producers to adopt better flock management and wool handling techniques—issues discussed in this and the previous chapter. In some ways, changes in pricing mechanisms so that prices are set according to clean yield and wool quality are fundamental to the success of any effort to improve wool production in China. Such changes, however, are unlikely to come quickly. Although there are standards for wool classing and grading implemented through the Fibre Testing Stations operating at prefectural levels under the overall leadership of the Standards Bureau in Xinjiang, for example, it can be assumed that such standards are implemented in a fairly rough-and-ready way throughout most of the basic purchasing agencies. Furthermore, until the herders have confidence in those standards and in the ability of the commercial agencies to pay a fair price for the wool bought, it is likely that they will continue to seek prices based on raw wool weight.

The implication of this situation is that both herders and commercial agencies have some vested interest in maintaining the present pricing system. At the very least, it will take time to develop appropriate standards and provide the practical means to implement them.

Nevertheless, the reforms over recent years have begun this process. The manager of Tianshan Mill reported that there is now a price differential of up to 10 per cent between wools of different qualities within the same grade. The fact that the market has also become a buyers' market means that textile enterprises are able to be more selective about wool quality and this will also encourage producers to be more careful in wool handling.

The decline in the quality of wool has, however, continued. The manager of Tianshan Mill reported that adulteration at the end of 1989 was still severe and that both clean yield and wool strength have been affected. He estimated that the usable wool per kg of clean wool has fallen from 80–85 per cent to 70 per cent. One of his workshops was devoted entirely to sorting out rubbish and sub-standard wool from the bales.

The other important issue in wool pricing is the relative pricing for wool and meat. Since wool is an industrial raw material, its pricing and marketing is sensitive to state intervention and subject to the pressures coming from industry. Meat prices are no longer controlled and the demand for meat is high. When these factors are combined with the high demand for mutton it is easy to see how sheep-raisers are very conscious of the relative prices offered for meat and wool and are likely to shape their breeding and management practices accordingly.

The above analysis suggests that, although reforms are underway, the structure of economic incentives in wool pricing and marketing systems are not shaped to promote the adoption of new breeding, flock management and wool handling techniques aimed at improving wool quality. Meanwhile, although the down-turn in the industry has, for the moment, removed some of the pressures that gave rise to the wool war of 1985–88, some of the underlying ingredients, especially in terms of competition between marketing systems operating to achieve different goals, remain present.

THE END OF THE WOOL WAR

As the above discussion has shown, the wool war came to an abrupt halt in 1989. The demand for raw wool collapsed, prices tumbled and producers were unable to sell their stocks. The end of the war, however, was not brought about by the emergence of an integrated market and the resumption of inter-regional flows. Nor did it result in improved capacity utilisation rates in the wool textile industry. Instead, it came as the result of a strict credit squeeze and a sharp drop in consumer demand.

The restrictions on capital supply during 1989 meant that both commercial departments and processors lacked the funds to purchase large volumes of wool. This was true of both state and local enterprises and of township enterprises. One report estimated that the latter accounted for around 300,000 of the 2 million or so spindles operating in 1987 (Men Xiuqi, 1990). By 1989 the proportion had probably increased but some 40 per cent of them ceased operating during the year (*Nongmin Ribao,* 22 March 1990: 2). Overall, therefore, there was a sharp decline in capacity utilisation rates.

The sharp decline in consumer demand during 1989 was experienced across all consumer goods markets. To some extent, it reflected the shift from panic buying in 1988 to increased saving, some of it enforced through the mandatory purchase of government bonds. The industrial slowdown also sharply curtailed bonus and other payments. During the first eleven months of 1989, domestic sales of worsted fabric and knitting wool dropped by 26 per cent and 25.5 per cent respectively (*Nongmin Ribao,* 22 March 1990: 2). Commercial departments rapidly ran down their stocks of wool products, and producers were forced to carry ever larger stocks of finished goods, further exacerbating their shortage or working capital.

Ultimately, however, the resolution of the war in this manner did not mean that the forces which caused the conflict had disappeared. The war was not a temporary phenomenon caused by the transition from a plan to a market regime. In effect, many of the fundamental features which had originally caused the conflict remained in place. Industrial capacity still exceeded the potential for raw wool supplies. The nature of local government ownership and budgetary relationships still embodied conflicts between regions. The coexistence of different marketing systems still implied continued competition for supplies. The easing of capital supply and a lift in consumer spending could easily spark off a new round of conflict. Furthermore, if the potential drop in sheep numbers and raw wool production that the economic forces of mid-1990 implied were realised, the next round of the cycle of boom and bust could be as intense as the preceding one.

The final resolution of the wool war thus requires not only a transition to an integrated market based on clearly defined wool standards but also changes in local government behaviour, in enterprise ownership and operation and in the trading relationship between producing areas and other regions.

CONCLUSION

The preceding sections began by reviewing in detail the production and marketing of raw wool in China up to 1985. The wool war broke out in that year, with 'battles' taking place to procure the raw material to supply the rapidly growing wool textile production capacity. These battles involved competition between various types of enterprises (planned and outside the plan), various levels of government and various kinds of merchants. They also affected the wool-growers, who were faced with new relative prices for agricultural products and new marketing methods. The conflict was ignited by a relaxation of central control of marketing at a time when regional economic interests became sharply identified. This led to a loss of controls over prices and subsequent efforts by the wool-producing areas themselves to adjust prices, impose sales requirements and restrict the movement of raw wool.

The wool war had its origins in the linkages between different aspects of the reforms in the rural sector. First, the responsibility for sheep management was transferred back to the households. This had some positive effects in terms of the management of the sheep owned by each family but it created problems in terms of the management of the total flock, in the provision of services to wool-growers, and in the use of pastures. Fragmentation appears to have raised some costs so that the net effect of the change was mixed.

Second, at the same time as the responsibility for sheep management was being transferred to the household, relative prices were also changing. The profitability of raising sheep for meat was rising more quickly than that for wool. This was because the reforms for meat marketing were proceeding more quickly than those for wool. So while households were becoming more sensitive to relative profitabilities, the signals they were receiving from the partial reforms up to that time were not in favour of wool production. Also, producers, who might have anticipated further reform in wool marketing and higher prices, were stockpiling wool.

Third, the wool textile industry was growing rapidly. The number of spindles was increasing both inside and outside the plan. The latter was prompted by the low price for raw wool, the rising demand for wool products in both the domestic and export markets, and the new autonomy for investment at local government level stimulated by the process of financial reform. However, the slow pace of wool marketing reform meant that the interests of the wool-buyers were not being communicated to producers. At the same time, the agencies who might have acted as intermediaries faced their own set of priorities and costs and were unwilling to perform that role. Furthermore, the expansion of spindle capacity in small local plants in the early 1980s in response to the economic signals of that period may have led to a decrease in the efficiency with which the wool is used. Indirectly, this also decreased the supply available to the rest of the industry.

Even more directly, the expansion of local processing in the wool-growing areas exacerbated the procurement problems facing the state processing sector in the coastal provinces, the response to this being greater reliance on imported wool. Thus China's position in world markets and the fluctuations in China's purchases are very much related to the sequence of domestic reforms. It is also likely, however, that there has been a structural change in China's position in the world market, and China's future purchases are likely to remain high.

Fourth, even though the reforms in wool pricing and marketing were slow, there were economic pressures on the system because of the scale of demand. Growers had the incentive to by-pass the state system and thereby undermine it. The state's response ultimately was to deregulate prices in 1985, but that unleashed all the pent-up forces. The reinstatement of controls at the local level and the progress of the wool war over the years 1985 to 1988, as detailed above, thus serve to underline the difficulties of setting the agenda for reform in an economy where different economic sectors are subject to different rates of change and where differing systems of ownership and management are operating side by side. Apart from the economic problems involved, this also reflects on the more sensitive political issues in the relationship between the national minority areas and the coastal regions, and between local and central levels of government.

The experience of the wool war has led to a new round of reforms of the marketing system. The most significant are the experiments underway to move to marketing of wool by auction where prices would be related to clean wool yields. The overall situation of wool supply and demand was changed dramatically in 1989 by the deflationary macroeconomic policies. At the same time, relative returns began to move in favour of meat compared to wool and a much higher slaughtering rate occurred over the 1989–90 winter. The ensuing drop in sheep numbers means that if fibre demand increases again, the situation will again become more of a sellers' rather than a buyers' market. If so, the competition between marketing systems, one subject to direction by the central planners and the other by regional interests, could re-emerge.

Appendix 4A

Construction of a consistent market price database

DAVID R. THOMPSON

This appendix describes the construction of a market price input–output database for the Chinese economy. Its purpose is to provide a detailed treatment of the data transformations as a reference for studies using this and related databases, and to complement the theoretical treatment presented in such papers as Martin (1990c). Basic data were taken from the World Bank (1985a) input–output (I–O) table. As the emphasis of the study was on the Chinese textile sector, a number of adjustments to the basic I–O table were undertaken to produce the required additional sectors. A range of sources have been drawn upon to make these adjustments, and some subjective adjustments were necessary to reconcile those sources.

The methods used to create the new sectors in the database will be discussed first, followed by an explanation of the price adjustments applied to this expanded I–O table. The process used to estimate the value added shares for each primary factor are then explained, and some alternative trade scenarios which involve a 'fine tuning' of the database are discussed.

ADDING THE COTTON, WOOL, TEXTILE AND APPAREL SECTORS

A cotton sector was derived from the crops sector by estimating a gross output value for cotton using the value of imports of cotton and the ratio of gross output to imports (Table 3.1). The cost structure of the cotton activity was first estimated using the *China Statistical Yearbook* (1984–85), making the assumption that in general the shares for all intermediate inputs and value adding factors were the same as those for the total crops category, and then splitting each cost component in proportion to total output value. Cotton imports were estimated using data from the International Economic Data Bank (IEDB) at the Australian National University, and converted to Chinese domestic currency, accounting for the exchange rate, tariff rate and exchange rate scarcity premium.

Approximately 90 per cent of the gross output value of cotton was allocated to the textile industry, the largest user of cotton as an input, with most of the remainder allocated to the chemical industry, a major user of crops output in the original I–O table. Cotton exports were based upon estimates in Table 3.1 and investment (stocks and fixed) were set at approximately 20 per cent of crop investment values to reflect the importance of stocks for a durable commodity such as cotton. It was assumed that the intermediate use of both crops and cotton output by the cotton sector was very small.

A wool sector was derived from the animal husbandry sector in the same manner as the cotton sector, applying the same assumptions, and using adjusted IEDB data to estimate wool imports. The ratio of production value to imports reported in Table 3.4 was then applied to estimate a gross output value for the wool sector. Judgement was used to allocate wool input use to industries which used animal husbandry inputs. In most instances, these allocations involved very small amounts, with the exception of the textile industry, which was assumed to use approximately 90 per cent of the gross output value of wool. Wool exports and sales to investment were set at zero.

The textile and clothing sector provided in the 1985 World Bank I–O table was divided into textile, apparel and chemical fibre sectors. Gross output of chemical fibres was calculated using the production to imports ratio for 1981 in Table 3.5 and an estimated value of imports in 1981. The value of imports was constructed by converting the value of imports in foreign currency to domestic currency using the exchange rate adjusted for a secondary market exchange rate premium.

Gross output in the chemical fibres sector was decomposed into its cost components using the most recent available estimate of the cost structure for this industry in a developing country. Using information on the Indonesian chemical fibres industry (Indonesia, Biro Pusat Statistik, 1984) with an adjustment to remove an unusually high input of hotel and restaurant services by this industry in Indonesia, it was estimated that 10 per cent of gross output was accounted for by intermediate inputs, and 90 per cent by gross value adding factors. Gross value added was further partitioned into depreciation, wages and profits, and taxes, which accounted for 2.7, 15.9 and 81.2 per cent respectively of value added. Intermediate input use was allocated over commodities using the same ratios reported in the Indonesian I–O table. Stock investment for chemical fibres was set at 0.25 billion yuan to reflect the importance of stock demand for these relatively durable commodities and to act as a balancing item. The total use of chemical fibres (8.4 billion yuan) was allocated to the textile industry, whose intermediate demand was assumed to be the only non-investment source of demand for chemical fibres.

The cost components of the chemical fibre sector were then subtracted from textile and clothing costs in the basic I–O table to give costs in a revised textile and clothing sector. The new textile clothing sector was then split into textiles and apparel, based initially upon information in the 1981 I–O table for China provided by the State Planning Commission and Statistical Bureau (1987), which made separate estimates for these sectors. Each of the intermediate input and value added figures were split using the shares implicit in this I–O table. The values of sales (namely demands) for the textile and apparel commodities were also based on information in the I–O table by the China State Planning Commission and Statistical Bureau.

FURTHER ADJUSTMENTS

At this stage, some further adjustments were made to the now expanded I–O table. First, the following alterations were made as explained by the World Bank (1985a: 51). These adjustments are largely to account for the pricing system for non-traded goods used in China:

1 the wage bill for the health and education sector was multiplied by 1.55;
2 the wage bill for public administration and defence was multiplied by 1.35; and
3 profits and taxes for the housing sector were increased by 2.75 billion yuan.

Second, the size of the apparel sector was doubled (in other words, all entries in the apparel column were multiplied by 2 so that the gross output value of the sector was doubled). This was carried out as it was believed that the size of the sector was substantially under-reported in the World Bank tables and the official data on which these were based. In particular, the production of apparel in township enterprises may contribute around 50 per cent of total production (Zhou, Dillon and Wan, 1988). On a judgemental basis, 10 billion yuan worth of textile inputs was also transferred from the textile industry to the apparel industry, as the use of textiles in the apparel sector initially appeared too small in comparison with that in the textile sector.

PRICE ADJUSTMENTS

As explained in Chapter 3, a dual price system operates in China involving both official and free-market prices. The modelling exercise in Chapter 4 assumes that production and consumption decisions are guided by price changes at the margin, so it was necessary to adjust the I–O table (which reflects official prices) to more closely resemble the free market situation. Peng Zhaoyang (University of Adelaide, pers. comm., 1988) generously provided data on the price relativities between planned and free-market prices for 100 products in seven major categories:

1 agricultural products;
2 textile and other light industrial products;
3 chemical products;
4 energy products;
5 building materials;
6 non-ferrous metals; and
7 metal products.

Two sets of price relativities were calculated. These were the ratio of 1988 plan prices to world price at the then prevailing official exchange rate of 3.72 yuan/US dollar, and the ratio of 1988 secondary market prices to world prices at a secondary market exchange rate of 5.7 yuan /US dollar. Peng (pers. comm., 1988) has noted that planned prices include procurement prices for agricultural products and ex-factory prices for industrial products. International or world market prices are those at major distributing centres, or f.o.b. or c.i.f. prices for some products in their major trading countries. The ratio of domestic plan to secondary market prices was calculated by first adjusting the two domestic/ foreign price ratios to a common exchange rate and then taking their ratio. An additional source of price adjustment factors for agricultural products was Lardy (1983b: 23) who reported that average rural market prices were 80 per cent above the quota price. The results of the above comparisons were used to construct a set of price adjustment factors which were applied to the I–O table so that it reflected free-market prices. The commodities for which Peng provided price relativities did not match exactly the commodities which appear in the I–O table. Therefore, the price adjustment factors were based upon the commodities from Peng's listing which would form the largest component of each I–O table commodity, with some adjustments where the reported price relativities appeared to be out of line with other sources. In general, China's domestic plan prices were considerably lower than international prices, so values in the I–O table were adjusted upwards. Table 4A.1 shows the price adjustment factors finally applied to each commodity.

This adjustment procedure reflects the observation that the official price system induces distortions in profitability (Chen Xikang, 1988). Typically, industries whose outputs are intermediate inputs face lower official prices and hence appear to have lower levels of profits. In some cases, this leads to paradoxical results in the official I–O table with the capital/labour ratio (as measured by profits and taxes divided by wages) being much higher, for example, in the textile sector than in the machinery sector.

As noted above, the gross output values changed following the price adjustment. The revised gross output values were transferred from the row sums to the column sums, with profits and taxes used as the balancing item. In several cases, the price adjustments led to apparent distortions in the returns to labour and capital for the service sectors. Some adjustments were made in the output prices for these relatively minor sectors to obtain realistic shares.

VALUE ADDING FACTORS

The final modification of the basic table was the estimation of primary factor shares such that factor value adding figures could be estimated. For the agricultural industries (crops, cotton, animal husbandry and wool), factor shares of 59 per cent for labour, 12 per cent for capital and 29 per cent for land implicit in McMillan, Whalley and Zhu (1987) were applied. This was necessary because of the high self-employment value for agriculture reported in the World Bank table. The self-employment value includes returns to labour, capital and

Table 4A.1 Price adjustments used in the revised Chinese I–O table

Commodity	Adjustment factor
Crops	1.80
Cotton	1.50
Animal husbandry	1.50
Wool	2.00
Metallurgy	1.60
Electricity	1.00
Coal	1.50
Petroleum mining	4.00
Petroleum refining	3.00
Chemicals	2.00
Chemical fibre	1.40
Machinery	1.40
Building materials	1.50
Wood	1.50
Food processing	1.40
Textiles	1.00
Apparel	1.00
Paper	1.10
Miscellaneous manufacturing	1.30
Construction	1.45
Freight transport	1.35
Passenger transport	1.28
Commerce	1.10
Miscellaneous services	1.00
Education and health	1.45
Public administration and defence	1.32
Housing	1.00

land, and thus it was necessary to obtain a set of shares from another source. All other industries were considered to use only labour and capital value adding factors, and the value of returns to capital was estimated as a balancing item following revision of the gross output value for each commodity obtained from the row sums of the table. The labour share of value added was estimated as wages divided by the revised value added after price adjustment. The capital share of value added was estimated as the sum of depreciation, interest and revised profits and taxes divided by revised value added. The final adjusted version of the 1981 I–O table for China is given in Table 4A.2.

Table 4A.3 shows labour shares in each industry for the original and adjusted I–O tables for comparative purposes. Because official prices for producer goods such as coal and machinery are particularly low in China (as outlined below), the profitability of these sectors is depressed and hence the labour share of primary factor returns is raised disproportionately. The adjustments made to the table typically lowered the labour share in producer goods industries such as metals, coal and machinery. Labour shares increased in some other industries which were not price adjusted (such as electricity) as prices of intermediate inputs were raised due to price adjustment in other sectors, and thus residual returns to capital declined.

ALTERNATIVE TRADE SCENARIOS

It was noted above that the trade structure within China has altered substantially since 1981. To cover this aspect, two alternative databases were constructed for use in the model. These will now be discussed.

1 Rest of world (ROW) export figures were estimated using IEDB data on China and ROW exports for 1981. The relative share of China's exports in ROW exports was used to scale the China export figures in the basic I–O table to produce a more accurate estimate of ROW exports for use in the model. The effect of this refined data set on model outcomes could then be gauged. The estimated China and ROW exports using the above method are shown in Table 4A.4.

2 A further database was constructed which incorporated a trade structure (in other words, an export and import structure) which resembled the more recent 1986 situation. IEDB data were used to calculate the share of exports and imports of each commodity in total Chinese exports and imports. A scaling factor was thereby created which would rescale 1981 exports and imports to a 1986 figure. A second scaling factor was also created which reflected the change in China's exports and imports as a share of GDP between 1981 and 1988 (World Bank, 1989b). The basic 1981 export–import figures were scaled by both of these factors. Hence a new 1986 trade structure was created which had the correct size in terms of share of GDP, and the correct commodity composition of exports and imports for 1986.

The new trade values were then price adjusted as explained in the preceding section. A revised gross output value was calculated as the row sum for each commodity. These revised values were then made the column totals (that is, industry gross outputs) in the I–O table, with profits and taxes used as the balancing item. The revised I–O table with a 1986 trade structure is given in Table 4A.5.

Table 4A.2 Input–output table for China, 1981 (market prices, current billion yuan)

Sector	Crops	Cotton	Animal husbandry	Wool	Metallurgy	Electricity	Coal	Petroleum Mining	Petroleum Refining
Crops	19.21	0.00	36.00	0.00	0.00	0.00	0.47	0.00	0.00
Cotton	0.00	0.00	0.00	0.00	0.00	0.00	0.06	0.00	0.00
Animal husbandry	0.00	0.00	1.29	0.21	0.00	0.00	0.00	0.00	0.00
Wool	0.00	0.00	0.00	0.00	0.00	0.00	0.00	0.00	0.00
Metallurgy	0.29	0.01	0.02	0.00	17.71	0.11	0.72	0.24	0.00
Electricity	1.22	0.19	0.29	0.05	2.81	2.87	1.22	0.35	0.17
Coal	0.88	0.14	0.21	0.03	3.00	4.34	1.65	0.02	0.00
Petrol mining	0.00	0.00	0.00	0.00	1.24	3.40	0.00	2.24	33.08
Petrol refining	7.07	1.15	1.68	0.27	1.68	2.16	0.36	0.54	0.60
Chemicals	25.02	3.98	0.02	0.00	0.60	0.02	0.50	0.80	0.20
Chemical fibre	0.00	0.00	0.00	0.00	0.00	0.00	0.00	0.00	0.00
Machinery	1.89	0.21	0.12	0.02	3.92	0.01	0.98	0.70	0.14
Building materials	0.20	0.02	0.06	0.02	0.15	0.02	0.08	0.15	0.02
Wood	0.14	0.02	0.06	0.02	0.38	0.02	0.90	0.02	0.02
Food processing	0.72	0.12	0.45	0.03	0.56	0.01	0.14	0.14	0.01
Textiles	0.24	0.04	0.06	0.01	1.04	0.01	0.50	0.70	0.03
Apparel	0.11	0.01	0.03	0.00	1.01	0.00	0.30	0.10	0.01
Paper	0.09	0.00	0.01	0.00	0.53	0.01	0.02	0.11	0.01
Miscellaneous manufacturing	0.12	0.01	0.06	0.01	0.52	0.01	0.26	0.13	0.01
Construction	0.66	0.07	0.00	0.00	2.61	0.00	2.32	0.00	0.00
Freight transport	0.87	0.14	0.06	0.01	3.20	0.12	0.41	0.41	0.11
Passenger transport	0.01	0.00	0.01	0.00	0.19	0.03	0.06	0.06	0.03
Commerce	1.06	0.07	0.05	0.01	1.58	0.10	0.40	0.57	0.02
Miscellaneous services	0.03	0.01	0.02	0.00	0.25	0.01	0.04	0.03	0.01
Education and health	0.00	0.00	0.00	0.00	0.00	0.00	0.00	0.00	0.00
Public administration and defence	0.00	0.00	0.00	0.00	0.00	0.00	0.00	0.00	0.00
Housing	0.00	0.00	0.00	0.00	0.00	0.00	0.00	0.00	0.00
Intermediate	59.82	6.18	40.49	0.69	42.98	13.25	11.38	7.30	34.46
Labour value added	137.05	8.45	11.97	0.28	3,90	1.09	5.78	0.73	0.10
Capital value added	27.87	1.72	2.43	0.06	26.28	4.48	7.82	43.57	10.29
Land value added	67.36	4.15	5.88	0.14	0.00	0.00	0.00	0.00	0.00
Gross output value	292.11	20.51	60.77	1.16	73.16	18.81	24.98	51.60	44.85

Table 4A.2 (continued)

Sector	Chemicals	Chemical fibre	Machinery	Building materials	Wood	Food processing	Textiles	Clothing	Paper
Crops	4.73	0.00	0.02	2.70	4.50	49.77	0.67	0.71	4.68
Cotton	1.31	0.00	0.00	0.00	0.00	0.00	16.62	0.18	0.00
Animal husbandry	0.05	0.00	0.01	0.00	0.00	16.96	1.19	0.03	0.00
Wool	0.04	0.00	0.01	0.00	0.00	0.0	1.74	0.00	0.00
Metallurgy	1.44	0.00	39.04	0.64	0.16	0.08	0.02	0.13	0.00
Electricity	3.36	0.05	1.40	0.71	0.14	0.44	0.94	0.05	0.43
Coal	2.07	0.00	0.78	2.49	0.15	0.65	0.70	0.08	0.33
Petrol mining	3.00	0.00	0.64	0.28	0.04	0.16	0.72	0.04	0.04
Petrol refining	7.02	0.16	2.46	0.48	0.27	1.17	2.00	0.12	0.39
Chemicals	31.86	0.02	5.20	1.00	0.50	5.60	3.58	1.22	3.00
Chemical fibre	0.00	0.00	0.00	0.00	0.00	0.00	11.76	0.00	0.00
Machinery	2.10	0.01	41.22	1.12	0.35	0.14	2.41	0.18	0.28
Building materials	0.30	0.00	0.45	2.27	0.15	0.08	0.07	0.00	0.00
Wood	0.08	0.00	0.60	0.60	2.07	0.02	0.28	0.03	1.95
Food processing	3.50	0.00	0.56	0.14	0.07	5.96	0.03	1.06	0.07
Textiles	1.47	0.01	1.00	0.80	0.14	0.21	31.92	35.00	0.40
Apparel	0.24	0.00	0.77	0.35	0.07	0.07	0.05	2.33	0.00
Paper	0.70	0.02	0.55	0.44	0.11	0.15	0.53	0.13	5.58
Miscellaneous manufacturing	0.59	0.07	0.85	0.26	0.09	0.01	1.21	0.23	0.26
Construction	1.09	0.00	2.18	0.87	0.00	0.00	0.00	0.00	0.00
Freight transport	1.96	0.14	4.66	1.22	0.45	1.11	0.00	0.84	0.54
Passenger transport	0.32	0.00	0.52	0.06	0.06	0.38	0.24	0.12	0.06
Commerce	2.90	0.02	3.52	2.06	0.24	0.79	2.06	0.24	0.54
Miscellaneous services	0.20	0.00	0.34	0.04	0.02	0.10	0.10	0.03	0.03
Education and health	0.00	0.03	0.00	0.00	0.00	0.00	0.00	0.00	0.00
Public administration and defence	0.00	0.00	0.00	0.00	0.00	0.00	0.00	0.00	0.00
Housing	0.00	0.00	0.00	0.00	0.00	0.00	0.00	0.00	0.00
Intermediate	70.32	0.54	106.77	18.52	9.58	83.85	78.83	42.76	18.58
Labour value added	4.52	0.56	14.27	4.47	2.68	3.73	4.85	4.06	2.14
Capital value added	48.36	2.93	37.32	11.82	4.22	1.20	23.26	2.25	1.20
Land value added	0.00	0.00	0.00	0.00	0.00	0.00	0.00	0.00	0.00
Gross output value	123.20	4.02	158.36	34.81	16.48	98.78	106.94	49.06	21.92

Table 4A.2 (continued)

Sector	Misc. m'facturing	Construction	Freight transport	Pass'ger transport	Commerce	Misc. services	Education & health	Public admin. & defence	Housing
Crops	3.78	6.30	0.00	0.00	6.66	0.00	1.86	0.00	0.00
Cotton	0.15	0.00	0.00	0.00	0.00	0.00	0.25	0.00	0.00
Animal husbandry	0.12	0.00	0.00	0.00	3.00	0.00	0.00	0.00	0.00
Wool	0.00	0.00	0.00	0.00	0.00	0.00	0.00	0.00	0.00
Metallurgy	1.28	10.37	0.8	0.00	0.00	0.00	0.00	0.00	0.00
Electricity	0.65	0.17	0.09	0.02	0.47	0.20	0.30	0.00	0.00
Coal	0.45	0.24	0.89	0.14	0.32	0.11	0.54	0.00	0.00
Petrol mining	0.32	0.00	0.00	0.00	0.00	0.00	0.00	0.00	0.00
Petrol refining	0.66	1.35	4.77	3.00	0.00	0.00	0.06	0.00	0.00
Chemicals	1.20	4.80	0.30	1.20	3.80	0.20	14.52	0.00	0.20
Chemical fibre	0.00	0.00	0.00	0.00	0.00	0.00	0.00	0.00	0.00
Machinery	0.42	13.58	1.26	1.44	6.02	0.21	2.24	0.00	0.21
Building materials	0.15	29.18	0.00	0.00	0.00	0.00	0.00	0.00	0.60
Wood	0.38	3.75	0.00	0.00	0.00	0.00	0.00	0.00	0.38
Food processing	0.70	0.14	0.14	0.00	5.60	0.28	2.10	0.00	0.00
Textiles	0.98	0.33	0.04	0.00	1.56	0.20	2.30	0.00	0.00
Apparel	0.27	1.18	0.16	0.00	0.14	0.10	1.10	0.00	0.00
Paper	0.44	0.50	0.22	0.00	2.20	0.33	5.17	0.00	0.00
Miscellaneous manufacturing	3.58	1.82	0.00	0.00	0.26	0.00	4.49	0.00	0.00
Construction	0.15	5.80	5.80	0.00	0.15	0.00	0.44	0.00	1.23
Freight transport	0.95	2.67	0.54	0.08	1.49	0.24	1.16	0.00	0.00
Passenger transport	0.06	0.20	0.06	0.00	0.13	0.00	0.00	0.00	0.00
Commerce	1.49	1.77	0.74	0.11	1.14	0.22	1.62	0.00	0.11
Miscellaneous services	0.03	0.08	0.06	0.00	0.10	0.10	0.35	0.00	0.00
Education and health	0.00	0.00	0.00	0.00	0.00	0.00	0.00	0.00	0.00
Public administration and defence	0.00	0.00	0.00	0.00	0.00	0.00	0.00	0.00	0.00
Housing	0.00	0.00	0.00	0.00	0.00	0.00	0.00	0.00	0.00
Intermediate	18.19	84.23	15.87	5.99	33.03	2.19	38.49	0.00	2.73
Labour value added	2.63	11.09	6.77	1.25	11.67	2.04	18.40	11.91	0.57
Capital value added	4.08	13.58	8.86	1.05	12.65	7.76	22.02	0.79	10.67
Land value added	0.00	0.00	0.00	0.00	0.00	0.00	0.00	0.00	0.00
Gross output value	24.90	108.90	31.50	8.29	57.35	11.99	78.91	12.70	13.97

Table 4A.2 (continued)

Sector	Intermediate	Consumption		Investment		Trade		Gross
		Household	Government	Fixed	Stocks	Export	Import	output
Crops	141.68	119.99	1.55	5.40	29.45	3.94	9.90	292.11
Cotton	18.75	0.00	0.00	1.11	6.06	0.07	5.30	20.51
Animal husbandry	22.85	32.25	0.00	0.00	4.47	1.20	0.00	60.77
Wool	1.79	0.00	0.00	0.00	0.00	0.00	0.63	1.16
Metallurgy	73.06	0.00	0.00	0.00	2.40	2.80	5.10	73.16
Electricity	18.59	0.35	0.50	0.00	0.00	0.00	0.00	18.81
Coal	20.18	4.04	0.41	0.00	0.00	0.50	0.14	24.98
Petrol mining	42.50	0.08	0.00	0.00	0.00	6.32	0.00	51.60
Petrol refining	39.42	0.99	0.87	0.00	0.00	3.63	0.06	44.85
Chemicals	109.34	11.94	2.00	0.00	3.12	5.92	9.12	123.20
Chemical fibre	11.76	0.00	0.00	0.00	0.35	0.00	6.47	4.03
Machinery	81.19	21.18	11.77	48.58	3.50	6.08	13.94	158.36
Building materials	33.94	0.00	0.30	0.00	0.30	0.29	0.02	34.81
Wood	11.67	4.97	0.75	0.00	0.30	0.57	1.77	16.48
Food processing	22.54	72.80	0.90	0.00	0.63	3.71	1.79	98.78
Textiles	78.99	20.59	0.52	0.00	3.74	6.00	2.90	106.94
Apparel	8.40	34.43	0.48	0.00	0.76	4.99	0.00	49.06
Paper	17.85	2.46	0.67	0.00	1.21	0.36	0.64	21.92
Miscellaneous manufacturing	14.84	4.04	2.60	0.00	1.82	3.64	2.04	24.90
Construction	23.35	0.00	3.05	82.51	0.00	0.00	0.00	108.90
Freight transport	23.36	3.81	0.16	1.35	1.08	3.31	1.57	31.50
Passenger transport	2.64	3.10	2.56	0.00	0.00	0.00	0.00	8.29
Commerce	23.43	28.53	0.30	2.20	0.72	3.78	1.61	57.35
Miscellaneous services	1.99	8.00	2.00	0.00	0.00	0.00	0.00	11.99
Education and health	0.03	3.60	75.28	0.00	0.00	0.00	0.00	78.91
Public administration and defence	0.00	0.00	12.70	0.00	0.00	0.00	0.00	12.70
Housing	0.00	13.97	0.00	0.00	0.00	0.00	0.00	13.97
Intermediate	847.00	391.12	119.36	141.15	59.90	57.11	62.99	1 550.03

Table 4A.3 Labour shares of value added in the original and adjusted I–O tables

Industry	1981 table from World Bank	Adjusted 1981 table
Crops	0.97[a]	0.59
Cotton	–	0.59
Animal husbandry	0.99[a]	0.59
Wool	–	0.59
Metallurgy	0.27	0.13
Electricity	0.10	0.19
Coal	0.79	0.42
Petrol mining	0.09	0.01
Petrol refining	0.02	0.09
Chemicals	0.24	0.09
Chemical fibre	–	0.16
Machinery	0.38	0.28
Building materials	0.46	0.27
Wood	0.57	0.39
Food processing	0.21	0.25
Textiles	0.24	0.17
Apparel	–	0.64
Paper	0.33	0.64
Miscellaneous manufacturing	0.44	0.39
Construction	0.61	0.45
Freight transport	0.54	0.43
Passenger transport	0.42	0.54
Commerce	0.41	0.48
Miscellaneous services	0.20	0.40
Education and health	0.99	0.93
Public administration and defence	0.99	0.94
Housing	0.10	0.05

Note: a Self-employment income is imputed to labour.

Table 4A.4 Exports for China and the rest of the world, 1981 (billion yuan)

Commodity	China exports	Rest of world exports
Crops	3.94	213.13
Cotton	0.07	3.48
Animal husbandry	1.20	40.01
Wool	–	393.00
Metallurgy	2.80	334.63
Electricity	–	2.64
Coal	0.50	36.24
Petrol mining	6.32	644.00
Petrol refining	3.63	306.43
Chemicals	5.92	558.18
Chemical fibre	–	4.42
Machinery	6.08	3961.50
Building materials	0.29	23.19
Wood	0.57	86.49
Food processing	3.71	607.64
Textiles	6.00	85.41
Apparel	4.99	88.13
Paper	0.36	77.57
Construction	–	–
Freight transport	3.31	162.07
Passenger transport	–	–
Commerce	3.78	185.42
Miscellaneous services	–	–
Education and health	–	–
Public administration and defence	–	–
Housing	–	–

Table 4A.5 Input–output table for China, 1986 (market prices, current billion yuan)

Sector	Crops	Cotton	Animal husbandry	Wool	Metallurgy	Electricity	Coal	Petroleum Mining	Petroleum Refining
Crops	19.21	0.00	36.00	0.00	0.00	0.00	0.47	0.00	0.00
Cotton	0.00	0.00	0.00	0.00	0.00	0.00	0.06	0.00	0.00
Animal husbandry	0.00	0.00	1.29	0.21	0.00	0.00	0.00	0.00	0.00
Wool	0.00	0.00	0.00	0.00	0.00	0.00	0.00	0.00	0.00
Metallurgy	0.29	0.01	0.02	0.00	17.71	0.11	0.72	0.24	0.00
Electricity	1.22	0.19	0.29	0.05	2.81	2.87	1.22	0.35	0.17
Coal	0.88	0.14	0.21	0.03	3.00	4.34	1.65	0.02	0.00
Petrol mining	0.00	0.00	0.00	0.00	1.24	3.40	0.00	2.24	33.08
Petrol refining	7.07	1.15	1.68	0.27	1.68	2.16	0.36	0.54	0.60
Chemicals	25.02	3.98	0.02	0.00	0.60	0.02	0.50	0.80	0.20
Chemical fibre	0.00	0.00	0.00	0.00	0.00	0.00	0.00	0.00	0.00
Machinery	1.89	0.21	0.12	0.02	3.92	0.01	0.98	0.70	0.14
Building materials	0.20	0.02	0.06	0.02	0.15	0.02	0.08	0.15	0.02
Wood	0.14	0.02	0.06	0.02	0.38	0.02	0.90	0.02	0.02
Food processing	0.72	0.12	0.45	0.03	0.56	0.01	0.14	0.14	0.01
Textiles	0.24	0.04	0.06	0.01	1.04	0.01	0.50	0.70	0.03
Apparel	0.11	0.01	0.03	0.00	1.01	0.00	0.30	0.10	0.01
Paper	0.09	0.00	0.01	0.00	0.53	0.01	0.02	0.11	0.01
Miscellaneous manufacturing	0.12	0.01	0.06	0.01	0.52	0.01	0.26	0.13	0.01
Construction	0.66	0.07	0.00	0.00	2.61	0.00	2.32	0.00	0.00
Freight transport	0.87	0.14	0.06	0.01	3.20	0.12	0.41	0.41	0.11
Passenger transport	0.01	0.00	0.01	0.00	0.19	0.03	0.06	0.06	0.03
Commerce	1.06	0.07	0.05	0.01	1.58	0.10	0.40	0.57	0.02
Miscellaneous services	0.03	0.01	0.02	0.00	0.25	0.01	0.04	0.03	0.01
Education and health	0.00	0.00	0.00	0.00	0.00	0.00	0.00	0.00	0.00
Public administration and defence	0.00	0.00	0.00	0.00	0.00	0.00	0.00	0.00	0.00
Housing	0.00	0.00	0.00	0.00	0.00	0.00	0.00	0.00	0.00
Intermediate	59.82	6.18	40.49	0.69	42.98	13.25	11.38	7.30	34.46
Labour value added	141.46	11.74	11.87	0.25	3,90	1.09	5.78	0.73	0.10
Capital value added	28.77	2.39	2.41	0.05	16.79	3.75	7.53	41.89	7.90
Land value added	69.53	5.77	5.84	0.12	0.00	0.00	0.00	0.00	0.00
Gross output value	299.59	26.09	60.61	1.11	63.67	18.08	24.69	49.92	42.47

Table 4A.5 (continued)

Sector	Chemicals	Chemical fibre	Machinery	Building materials	Wood	Food processing	Textiles	Clothing	Paper
Crops	4.73	0.00	0.02	2.70	4.50	49.77	0.67	0.71	4.68
Cotton	1.31	0.00	0.00	0.00	0.00	0.00	16.62	0.18	0.00
Animal husbandry	0.05	0.00	0.01	0.00	0.00	16.96	1.19	0.03	0.00
Wool	0.04	0.00	0.01	0.00	0.00	0.0	2.47	0.00	0.00
Metallurgy	1.44	0.00	39.04	0.64	0.16	0.08	0.02	0.13	0.00
Electricity	3.36	0.05	1.40	0.71	0.14	0.44	0.94	0.05	0.43
Coal	2.07	0.00	0.78	2.49	0.15	0.65	0.70	0.08	0.33
Petrol mining	3.00	0.00	0.64	0.28	0.04	0.16	0.72	0.04	0.04
Petrol refining	7.02	0.16	2.46	0.48	0.27	1.17	2.00	0.12	0.39
Chemicals	31.86	0.02	5.20	1.00	0.50	5.60	3.58	1.22	3.00
Chemical fibre	0.00	0.00	0.00	0.00	0.00	0.00	11.76	0.00	0.00
Machinery	2.10	0.01	41.22	1.12	0.35	0.14	2.41	0.18	0.28
Building materials	0.30	0.00	0.45	2.27	0.15	0.08	0.07	0.00	0.00
Wood	0.08	0.00	0.60	0.60	2.07	0.02	0.28	0.03	1.95
Food processing	3.50	0.00	0.56	0.14	0.07	5.96	0.03	1.06	0.07
Textiles	1.47	0.01	1.00	0.80	0.14	0.21	31.92	35.00	0.40
Apparel	0.24	0.00	0.77	0.35	0.07	0.07	0.05	2.33	0.00
Paper	0.70	0.02	0.55	0.44	0.11	0.15	0.53	0.13	5.58
Miscellaneous manufacturing	0.59	0.07	0.85	0.26	0.09	0.01	1.21	0.23	0.26
Construction	1.09	0.00	2.18	0.87	0.00	0.00	0.00	0.00	0.00
Freight transport	1.96	0.14	4.66	1.22	0.45	1.11	0.00	0.84	0.54
Passenger transport	0.32	0.00	0.52	0.06	0.06	0.38	0.24	0.12	0.06
Commerce	2.90	0.02	3.52	2.06	0.24	0.79	2.06	0.24	0.54
Miscellaneous services	0.20	0.00	0.34	0.04	0.02	0.10	0.10	0.03	0.03
Education and health	0.00	0.03	0.00	0.00	0.00	0.00	0.00	0.00	0.00
Public administration and defence	0.00	0.00	0.00	0.00	0.00	0.00	0.00	0.00	0.00
Housing	0.00	0.00	0.00	0.00	0.00	0.00	0.00	0.00	0.00
Intermediate	70.32	0.54	106.77	18.52	9.58	83.85	79.56	42.76	18.58
Labour value added	4.52	0.56	14.27	4.47	2.68	3.73	4.85	4.06	2.14
Capital value added	45.64	2.93	16.26	11.71	2.60	4.35	12.44	2.25	1.01
Land value added	0.00	0.00	0.00	0.00	0.00	0.00	0.00	0.00	0.00
Gross output value	120.48	4.02	137.30	34.70	14.86	91.92	96.85	49.06	21.73

Table 4A.5 (continued)

Sector	Misc. manufacture	Construction	Freight transport	Pass'ger transport	Commerce	Misc. services	Education & health	Public admin. & defence	Housing
Crops	3.78	6.30	0.00	0.00	6.66	0.00	1.86	0.00	0.00
Cotton	0.15	0.00	0.00	0.00	0.00	0.00	0.25	0.00	0.00
Animal husbandry	0.12	0.00	0.00	0.00	3.00	0.00	0.00	0.00	0.00
Wool	0.00	0.00	0.00	0.00	0.00	0.00	0.00	0.00	0.00
Metallurgy	1.28	10.37	0.8	0.00	0.00	0.00	0.00	0.00	0.00
Electricity	0.65	0.17	0.09	0.02	0.47	0.20	0.30	0.00	0.00
Coal	0.45	0.24	0.89	0.14	0.32	0.11	0.54	0.00	0.00
Petrol mining	0.32	0.00	0.00	0.00	0.00	0.00	0.00	0.00	0.00
Petrol refining	0.66	1.35	4.77	3.00	0.00	0.00	0.06	0.00	0.00
Chemicals	1.20	4.80	0.30	1.20	3.80	0.20	14.52	0.00	0.20
Chemical fibre	0.00	0.00	0.00	0.00	0.00	0.00	0.00	0.00	0.00
Machinery	0.42	13.58	1.26	1.44	6.02	0.21	2.24	0.00	0.21
Building materials	0.15	29.18	0.00	0.00	0.00	0.00	0.00	0.00	0.60
Wood	0.38	3.75	0.00	0.00	0.00	0.00	0.00	0.00	0.38
Food processing	0.70	0.14	0.14	0.00	5.60	0.28	2.10	0.00	0.00
Textiles	0.98	0.33	0.04	0.00	1.56	0.20	2.30	0.00	0.00
Apparel	0.27	1.18	0.16	0.00	0.14	0.10	1.10	0.00	0.00
Paper	0.44	0.50	0.22	0.00	2.20	0.33	5.17	0.00	0.00
Miscellaneous manufacturing	3.58	1.82	0.00	0.00	0.26	0.00	4.49	0.00	0.00
Construction	0.15	5.80	5.80	0.00	0.15	0.00	0.44	0.00	1.23
Freight transport	0.95	2.67	0.54	0.08	1.49	0.24	1.16	0.00	0.00
Passenger transport	0.06	0.20	0.06	0.00	0.13	0.00	0.00	0.00	0.00
Commerce	1.49	1.77	0.74	0.11	1.14	0.22	1.62	0.00	0.11
Miscellaneous services	0.03	0.08	0.06	0.00	0.10	0.10	0.35	0.00	0.00
Education and health	0.00	0.00	0.00	0.00	0.00	0.00	0.00	0.00	0.00
Public administration and defence	0.00	0.00	0.00	0.00	0.00	0.00	0.00	0.00	0.00
Housing	0.00	0.00	0.00	0.00	0.00	0.00	0.00	0.00	0.00
Intermediate	18.19	84.23	15.87	5.99	33.03	2.19	38.49	0.00	2.73
Labour value added	2.63	11.09	6.77	1.25	11.67	2.04	18.40	11.91	0.57
Capital value added	5.25	13.58	10.70	1.05	14.57	7.76	22.02	0.79	10.67
Land value added	0.00	0.00	0.00	0.00	0.00	0.00	0.00	0.00	0.00
Gross output value	26.07	108.90	33.34	8.29	59.27	11.99	78.91	12.70	13.97

Table 4A.5 (continued)

Sector	Intermediate	Consumption		Investment		Trade		Gross
		Household	Government	Fixed	Stocks	Export	Import	output
Crops	141.68	119.99	1.55	5.40	29.45	4.51	2.98	299.59
Cotton	18.57	0.00	0.00	1.11	6.06	0.36	0.01	26.09
Animal husbandry	22.85	32.25	0.00	0.00	4.47	1.04	0.00	60.61
Wool	2.47	0.00	0.00	0.00	0.00	0.00	1.36	1.11
Metallurgy	73.06	0.00	0.00	0.00	2.40	1.63	13.43	63.67
Electricity	18.59	0.35	0.50	0.00	0.00	0.00	0.00	18.08
Coal	20.18	4.04	0.41	0.00	0.00	0.45	0.37	24.69
Petrol mining	45..20	0.08	0.00	0.00	0.00	4.64	0.00	49.92
Petrol refining	39.42	0.99	0.87	0.00	0.00	1.68	0.50	42.47
Chemicals	109.34	11.94	2.00	0.00	3.12	5.89	11.81	120.48
Chemical fibre	11.76	0.00	0.00	0.00	0.35	0.00	2.98	4.03
Machinery	81.19	21.18	11.77	48.58	3.50	9.67	38.60	137.30
Building materials	33.94	0.00	0.30	0.00	0.30	0.19	0.03	34.70
Wood	11.67	4.97	0.75	0.00	0.30	0.47	3.29	14.86
Food processing	22.54	72.80	0.90	0.00	0.63	4.11	9.06	91.92
Textiles	78.99	20.59	0.52	0.00	3.74	4.21	11.20	96.85
Apparel	8.40	26.16	0.48	0.00	0.76	9.91	0.00	49.06
Paper	17.85	2.46	0.67	0.00	1.21	0.33	0.80	21.73
Miscellaneous manufacturing	14.84	4.04	2.60	0.00	1.82	8.42	5.65	26.07
Construction	23.35	0.00	3.05	82.51	0.00	0.00	0.00	108.90
Freight transport	23.36	3.81	0.16	1.35	1.08	3.58	0.00	33.34
Passenger transport	2.64	3.10	2.56	0.00	0.00	0.00	0.00	8.29
Commerce	23.43	28.53	0.30	2.20	0.72	4.10	0.00	59.27
Miscellaneous services	1.99	8.00	2.00	0.00	0.00	0.00	0.00	11.99
Education and health	0.03	3.60	75.28	0.00	0.00	0.00	0.00	78.91
Public administration and defence	0.00	0.00	12.70	0.00	0.00	0.00	0.00	12.70
Housing	0.00	13.97	0.00	0.00	0.00	0.00	0.00	13.97
Intermediate	847.73	382.85	119.36	141.15	59.90	65.20	102.06	1510.58

Appendix 4B

Equations of the model and data sources

WILL MARTIN

The set of equations making up the model is presented in Table 4B.1 together with the definitions of the variables and coefficients. The first six sets of equations specify the final demands for goods and the demands for intermediate goods by each sector. The first set of equations specifies the demand for each good by households as a function of household disposable income and the (marginal) prices of each good. In the absence of any elasticity estimates based on the appropriate marginal prices, the set of elasticities was calculated using expenditure elasticities for each good (World Bank, 1985b), budget share data at market prices, and an estimate of the Frisch Parameter (-6.9) obtained by interpolating from the international estimates made by Lluch, Powell and Williams (1977). It was necessary to slightly rescale all of the expenditure elasticities following the price adjustment in order to preserve the lingel aggregation property (Dervis, de Melo and Robinson, 1981: 482–5). The resulting estimates satisfy all the theoretical constraints such as homogeneity of degree zero in prices and income, symmetry of the compensated cross price effects and adding up.

Equation categories (2), (3) and (4) specify proportional changes in fixed investment, stocks investment and government consumption demands for each commodity as equal to the proportional changes in gross real absorption in the economy. This specification is seen as an essentially 'neutral' approach to specifying components of demand whose behaviour is not central to the analysis and which are subject to substantial, exogenous, government policy control.

Equation category (5) represents China's external trade environment. The demand for China's export of each commodity i is represented using a CES function (linearised in percentage changes) consistent with the Armington (1969) model. The demand for exports of good i is determined by the relative prices of China's exports and exports from the rest of the world, and the total demand for that particular commodity.

Table 4B.1 Model equations and variables

Equation	Number
1. Household consumption demands	
$q_i^{(3)} = \varepsilon_i a^* + \Sigma_{k=1}^{g} \eta_{ik} p_k^q$	g
2. Fixed investment demand	
$q_i^{(2)} = a_R$	g
3. Investment in stocks	
$qs_i^{(2)} = a_R$	g
4. Government demand	
$q_i^{(5)} = a_R$	g
5. Traded good demand/supply	
(a) Export demand from China	
$q_i^{(4)} = qw_i^{(4)} - \sigma_i^w (p_{is}^e - \Sigma_{s=1}^2 ES_{is} p_{is}^e)$	g
(b) World demand	
$qw_i^{(4)} = \beta_i (\Sigma_{s=1}^2 ES_{is} p_{is}^e)$	g
(s=1, China; 2, rest of world)	
(c) Import supply to China	
$q_{is} = E_i p_i^m$	g
6. Intermediate demands	
(a) Excluding textile sector	
$q_{ij}^{(1)} = x_j$	g^2-g
(b) Textile sector	
$q_{kt}^{(1)} = x_t - \sigma_t^k (p_k - \Sigma_j F_j p_j)$	g
7. Domestic absorption of good i from all sources	
$q_i = \Sigma_j B_{ij}^{(1)} q_{ij}^{(1)} + B_i^{(2)} q_i^{(2)} + BS_i^{(2)} qs_i^{(2)} + B_i^{(3)} q_i^{(3)} + B_i^{(5)} q_i^{(5)}$	g
8. Domestic import substitution	
$q_{is} = q_i - \sigma_i^m (p_{is} - p_i^q)$	$2g$
(s=1, imported; 2, domestic)	
9. Transformation in production	
$x_{id} = x_i + \sigma_i^T (p_{id} - p_i^x)$	$2g$
(s=1, export; 2, domestic)	
10. Primary factor inputs	
$q_{vj}^p = x_j - \sigma_i^p (p_{vj}^p - \Sigma_{v=1}^3 S_{vj}^p p_{vj}^p)$	$3g$

Table 4B.1 (continued)

11. <u>Product market clearing</u>

 (a) Domestic market clearing

 $q_{i2} = x_{i2}$ — g

 (b) Export market clearing

 $q_i^{(4)} = x_{i1}$ — g

12. <u>Factor market clearing</u>

 (a) $q_1^p = \Sigma_{j=1}^g L_j q_{1j}^p$ - Labour — 1

 (b) $q_{2j}^p = k_j$ - Capital in i — g

 (c) $q_{3j}^p = l_j$ - Land in i — g

13. <u>Zero pure profits at the margin</u>

 (a) In production

 $\Sigma_{d=1}^2 J_{jd} p_{jd} = \Sigma_i H_{ij}^{(1)} p_i^q + \Sigma_{v=1}^3 H_{vj}^p p_{vj}^p$ — g

 (b) In importing

 $p_{i2} = p_i^m + t_i + \phi_2$ — g

 (c) In exporting

 $p_{i1} = p_i^e + v_i + FES_1(RC{*}rr_i + \phi_1)$
 $+ (1-FES_1)(rr_i + \phi_2)$ — g

14. <u>GDP, absorption and household absorption</u>

 (a) $gdp_r = \Sigma_i K_i x_i$ — 1

 (b) $gdp = \Sigma_i K_i(p_i^x + x_i)$ — 1

 (c) $a_R = \Sigma_i SN_{i3} q_i^{(3)} + \Sigma_i SN_{i5} q_i^{(5)}$
 $+ \Sigma_i SN_{i2} q_i^{(2)} + \Sigma_i SN_{i6} qs_i^{(2)}$ — 1

 (d) $a = \Sigma_i SN_i a{*} + \Sigma_i SN_{i5}(p_i^q + q_i^{(5)})$
 $+ \Sigma_i SN_{i2}(p_i^q + q_i^{(2)}) + \Sigma_i SN_{i6}(p_i^q + qs_i^{(2)})$ — 1

15. <u>Balance of trade condition</u>

 $\pi = SXe + SMm$ — 1

16. <u>Balance of trade identities</u>

 (a) Total export value

 $e = \Sigma_i V_i (p_{i1} + x_{i1})$ — 1

Table 4B.1 (continued)

(b) Total import value

$m = \Sigma_i M_i (p_{i2} + q_{i1})$ — 1

(c) Total export volume

$e_R = \Sigma_i V_i x_{i1}$ — 1

(d) Total import volume

$m_R = \Sigma_i M_i q_{i1}$ — 1

17. <u>Composite price variables</u>

(a) Price level determination

$p^q = ms - a_R$ — 1

(b) Price deflator for *gdp*

$p^x = \Sigma_i K_i p_i^x$ — 1

(c) Price deflator for total absorption

$p^q = \Sigma_i W_i p_i^q$ — 1

(d) Price deflator for absorption of *i*

$p_i^q = \Sigma_{s=1}^2 A_{is} p_{is}$ — g

(e) Price deflator for output of *i*

$p_i^x = \Sigma_{d=1}^2 J_{id} p_{id}$ — g

Total nº of equations	$g^2 + 24g + 13$

<u>Exogenous variables (percentage change)</u>

a_R = Real absorption

k_j = Capital stock in industry *j*

l_j = Land use by industry *j*

ms = Money supply

p_{is}^e = Foreign currency price of good *i*, *s*=2, ROW

q_1^p = Total labour force

rr_i = Foreign exchange retention rate for exports of *i*

t_i = Power of the tariff on imports of *i* (1 + nominal tariff rate)

v_i = Power of the export tax on exports of *i* (1 - nominal export tax)

ϕ_1 = Official exchange rate (yuan/US dollar)

Table 4B.1 (continued)

<u>Value share coefficients</u>

A_{is} = Share of absorption of i derived from source s

$B_{ij}^{(1)}$ = Share of intermediate use of j in total absorption of i

$B_{i}^{(2)}$ = Share of investment in total absorption of commodity i

$BS_{i}^{(2)}$ = Share of stock demand in total absorption of commodity i

$B_{i}^{(3)}$ = Share of household consumption in total absorption of commodity i

$B_{i}^{(5)}$ = Share of government in total absorption of i

ES_{is} = Share of China and ROW in world export markets for i

F_{j} = Share of fibre j in total fibre demand by the textile industry

FES_{1} = Share of export revenue obtained from sales at official exchange rate,
$((1-R_{o})\phi_1)/(R_{o}\phi_2+(1-R_{o})\phi_1)$
where R_{o} = base period retention rate

$H_{ij}^{(1)}$ = Share of intermediate good i in total costs of industry j

H_{vj}^{p} = Share of primary factor v in total costs of industry j

J_{id} = Share of good i production to destination 1, export; 2, domestic

K_{i} = Share of sector i in total value added

L_{j} = Share of industry j in total employment

M_{i} = Share of i in total imports

RC = Conversion factor from proportional change in retention rate (R) to change in (1 - R), i.e. $(-R_{o}/(1-R))$

S_{vj}^{p} = Share of primary factor v in primary factor inputs of j

SM = Imports as a share of nominal *gdp*

SN_{ij} = Share of end-use demand j for commodity i in final absorption

SX = Exports as a share of nominal *gdp*

V_{i} = Share of i in total exports

W_{i} = Share of good i in total absorption

Table 4B.1 (continued)

Elasticity parameters

β_i	=	Global elasticity of excess demand for good *i*
E_i	=	Elasticity of import supply for good *i* to China
E_i	=	Household expenditure elasticity for good *i*
η_{ij}	=	Price elasticity of household demand for good *i* with respect to price *j*
σ_t^k	=	Elasticity of substitution between fibres in the textile industry (set to zero for non-fibre inputs)
σ_i^m	=	Elasticity of substitution between import and domestic products of good *i*
σ_i^p	=	Elasticity of substitution between primary factor inputs in sector *i*
σ_i^T	=	Elasticity of transformation between domestic and export production of good *i*
σ_i^W	=	Elasticity of substitution between Chinese and ROW products in world market for *i*

Endogenous variables (percentage change)			Nº
a	-	Nominal absorption	1
$a*$	-	Household nominal absorption	1
e	-	Export value	1
e_R	-	Export volume	1
gdp_r	-	Real *gdp*	1
gdp	-	Nominal *gdp*	1
m	-	Import value	1
m_R	-	Import volume	1
p_{is}^e	-	Foreign currency price of export *i*, *s*=1, China	g
p_{vj}^p	-	Return to primary factor *v* in industry *i*	$2g + 1$
p_i^m	-	Foreign currency price of import *i*	g
p^q	-	Composite price for absorption	1
p_i^q	-	Price for absorption of *i*	g
p_{ik}	-	Price of *i* in 1, export; 2, import; 3, domestic	$3g$

Table 4B.1 (continued)

Variable		Description	Number
p_i^x	-	Price for production of *i* (composite of domestic and export)	g
p^x	-	Aggregate price of output (*gdp* deflator)	1
π	-	Balance of trade as a share of *gdp*	1
q_i	-	Total absorption of *i*	g
$q_{ij}^{(1)}$	-	Intermediate use of *i* by industry *j*	g
$q_i^{(2)}, qs_i^{(2)}$	-	Stock demand for good *i*	$2g$
$q_i^{(3)}$	-	Household demand for *i*	g
$q_i^{(4)}$	-	Export demand for *i* from China	g
$qw_i^{(4)}$	-	World demand for good *i*	g
$q_i^{(5)}$	-	Government demand for *i*	g
q_{is}	-	Demand for *i* from source s=1, import; 2, domestic	$2g$
q_{vj}^p	-	Demand for primary factor *v* by industry *j* (v=1, labour; 2, capital; 3, land)	$3g$
x_j	-	Output level of industry *j*	g
x_{id}	-	Supply of good *i* to destination d; export (1) or domestic (2)	$2g$
ϕ_2	-	Secondary market exchange rate	1
Total nº of endogenous variables			$g^2 + 24g + 13$

In turn, the world demand was specified as a linear (in proportional changes) function of the weighted average price for good i where the weights were the shares of China and the rest of the world in total exports of good i. The supply of imports is specified as a function of the world price of imports, with increases in the foreign currency price of imports increasing the supply.

Equation set (6) specifies the demands for intermediate inputs of commodities in the production process. For simplicity, and for consistency with most models of this type, intermediate inputs are generally assumed to be used in fixed proportion to outputs; that is, according to a Leontief technology. The exception to this is the demand for fibres by the textile industry. Here it is assumed that the three fibres identified in the model — wool, cotton and chemical fibres — are substitutes, and that the substitution between these fibres can be approximated by a CES function.

Equation (7) aggregates intermediate usage, household stocks, consumption and government demand into a total absorption variable. Value share weights are used to convert this linear identity into percentage change form. Because export and domestic products are differentiated in this model, export demands are not a component of total absorption of i.

Equation set (8) specifies substitution in demand between good i from domestic and imported sources. These equations are a linear (in percentage changes) version of the Armington (1969) specification of imperfect substitution between domestic and imported products.

Equation set (9) specifies imperfect transformation between domestically produced products supplied to domestic and export markets. This equation is a linearisation in percentage changes of the Constant Elasticity of Transformation (CET) function originated by Powell and Gruen (1968) and discussed in the context of CGE models by Robinson (1989). The justification for this assumption is that the products supplied to the export market are to some extent differentiated from those supplied to the domestic market, either in terms of tangible attributes such as design and manufacture, or by intangible differences such as different market requirements.

Equation set (10) specifies the demand for primary factor inputs by industry i as a function of the output level in industry i and the relative prices of each of the primary factor inputs (land, labour and capital). It is assumed that these inputs can be aggregated into a composite primary factor bundle using a CES function, and the demand equations are obtained by imposing the first order conditions for cost minimisation, and linearising in percentage changes.

The market clearing conditions for commodities are specified in equation block (11). In 11(a) domestic demand for good i from domestic sources (q_{i1}) is equated with domestic production of good for the home market (x_{i1}). Similarly, export demand for good i from China must equal Chinese production of good i for export (x_{i1}).

Equation set (12) deals with market clearing for primary factors. Equation 12(a) embodies the assumption, standard in models of this type, that labour is able to move between different industries in response to changes in demand for labour. While this is undoubtedly a strong assumption given the restrictions on the physical mobility of labour in China, the explosive growth of the lightly regulated rural industries in China has greatly increased the opportunities for labour to move between agriculture and industry, and between industries. Furthermore, there appears to have been a tendency for restrictions on the physical movement of labour to be relaxed over time.

The stock of capital in each industry, and the stock of land in each agricultural industry, is assumed to be fixed during the period under consideration. While investment in fixed capital was considered as a component of final demands, this investment is assumed to be unable to affect the effective capital stock. This assumption makes the model effectively short-run in nature, with a time horizon of perhaps around two years. This short-run analysis can be thought of as providing an indication of the pressures for change over the longer term.

Equation set (13) imposes the condition of zero pure profits on activities conducted at marginal (free-market) prices. In production, this condition involves the inherent assumption of constant returns to scale. Given the very large number of enterprises involved in most industrial (and certainly agricultural) activities in China, this assumption appears reasonable. While the two-tier pricing system generates large profits and losses, these are assumed to be infra-marginal and hence are ignored given the intended focus of this model on resource allocation issues rather than distributional issues.

The zero-profit or arbitrage conditions in exporting and importing are of central importance and are therefore examined in some detail. The condition for the import market is simply a linear in percentage change version of

$$P_{i2} = P^{m}_{i2} (1+T_i) e_2$$

where P_{i2} is the landed, domestic currency price of imported good i, T_i is the rate of tariff applying to imports of i (or the tariff equivalent of the quota if imports of i are restricted by quantitative controls), P^{m}_{i2} is the c.i.f. price of imports in foreign currency, and e_2 is the nominal exchange rate in the secondary market for foreign exchange. At the margin, it is assumed that the opportunity cost of all imports involves the secondary market rate. If an enterprise has less foreign exchange than it demands, it must purchase additional foreign exchange in the secondary market. If it has more foreign exchange initially than it requires, its opportunity cost of using foreign exchange is also the secondary market rate.

In exporting, the returns available depend upon the foreign currency price received, the rate of any export tax, and a weighted average of the official and secondary market exchange rates. The higher the rate of retention allowed to enterprises, the larger is the weight on the secondary market, and hence the higher the domestic currency price of exports. Equation 13(c) was derived by expanding the non-linear export value equation about base period values for the retention rate and the official and secondary market exchange rates. It allows the effects of changes in both the official exchange rate and in the retention rate applied in each industry. The weight FES reflects an assumed base retention rate to enterprises of 0.25 and allows the effects of changes from this level to be investigated.

Equation set (14) includes identities to form aggregate GDP and absorption in current and constant prices. Equation 14(d) requires that total household absorption, and the spending associated with investment and government purchases, add to total absorption. Equation (15) defines the balance of trade as a proportion of GDP depending upon the relationship between income and expenditure. This presupposes the existence of some mechanism by which total income and total expenditure can be kept broadly in line and hence the current account can be broadly controlled. This simple equation summarises the consequences of a broad range of consumption and investment decisions and government fiscal and expenditure control variables which it would not be realistic to model in this study, particularly given the apparent complexity of fiscal and taxation policy in contemporary China (Blejer and Szapary, 1989). Given the focus of this study on issues of trade and resource allocation, this approach seems adequate.

Equation 16(a) is a demand for money equation, with a unitary income elasticity imposed. This equation essentially imposes the quantity theory of money and implies that a 1 per cent increase in the money supply will raise the GDP price deflator by 1 per cent. Chow (1987) finds evidence that the quantity theory may apply reasonably well to China and in fact concluded, somewhat surprisingly, that it appeared to offer a better representation of price behaviour for China than for the United States.

While Chow's results did not support the strict quantity theory result that increase in the money supply would have an equiproportional effect on the price level, this result may have reflected the rigidity of official prices over much of his sample period. Feltenstein and Ziba (1987: 153) conclude that the official price indexes of the type used by Chow substantially understated the true rate of inflation. Combining this estimate, that the true rate of inflation in China is 2.5 times the official rate, with the Chow result, that a 1 per cent increase in the money supply would raise official prices by one-third of 1 per cent, suggests that the unit elasticity used in this study may not be unreasonable. The evolution of institutions for management of monetary policy (de Wulf and Goldsborough, 1986) provides at least the potential for policy control of the money supply.

The final four equations of the model 17(b)–17(e) define the price of each of the composite goods (import–domestic) consumed domestically, and the composite price (export–domestic) good produced domestically and composite prices for total absorption and total GDP.

The complete model contains 1,390 equations and 1,390 endogenous variables. The model is linear in percentage changes and was solved using the GEMPACK program (Codsi and Pearson, 1988).

In addition to the input–output data discussed above, the model requires that a number of elasticity parameters be specified. The elasticities involved were:

1 consumer expenditure elasticities;
2 elasticities of substitution between fibres in the textile sector (base value 1.0);
3 elasticities of substitution between domestic and imported good i (base value 2.0);
4 elasticities of transformation between domestic and exported good i (base value 5.0);
5 elasticities of substitution between Chinese exports of i and the exports of

other countries (base value 10.0);
6 the elasticity of demand for total world exports of i (base value -2.0);
7 the elasticity of supply of import i to China (base value 100); and
8 the elasticity of substitution between primary factors in industry i (base value 0.5).

As previously discussed, the consumer demand elasticities were derived using expenditure elasticities, budget shares and the Frisch Parameter. In most cases, the absolute values of the own-price elasticities were relatively small, with many in the range of -0.2 to -0.3. As is common with this approach, the cross price elasticities are generally negative, with the negative income effects of price rises outweighing the positive substitution effects.

It seemed unlikely that reasonable estimates of the elasticities of substitution and transformation could be estimated using the available data for China. Time series of the relevant market price data are extremely scarce and, in any event, the time period over which enterprise managers have been free to allocate their resources in response to relative price changes is relatively short. Accordingly, the approach taken was to impose selected base values chosen on the basis of overseas experience and then to examine the sensitivity of the results obtained to these assumptions.

The base value of 1.0 for the elasticity of substitution between fibres in the production of textiles is in the same order of magnitude as the recent estimates of fibre demand elasticities reported by Ball, Beare and Harris (1989). While the 'true' elasticity of substitution may be somewhat lower, particularly in the short run, the limited available econometric evidence on the elasticity of fibre demand (see, for example, Hinchy and Fisher, 1988) makes it difficult to support the hypothesis of a substantially higher value.

The base value of 2.0 used for the elasticity of substitution between domestic and imported commodities is within the range of values used for the parameter in CGE studies. While the values used in the Grais, de Melo and Urata (1986) study range only from 0.4 to 1.2, the corresponding parameters are frequently larger in many other CGE studies. If one accepts the weight of empirical evidence marshalled by Goldstein and Khan (1985: 1,076) that the aggregate elasticity of import demand is in the range -0.5 to -1.0, then an elasticity of substitution of 2.0 at the individual commodity level would seem entirely reasonable.

Unfortunately, the empirical evidence on the elasticity of transformation between domestic and export production is extremely limited. The estimate of 2.90 cited by Tarr (1989: 5–7) perhaps provides some indication of the order of magnitude, at least for manufactured products. While well below the value of infinity implicit in models constructed without explicit transformation in production, it is well above the values of 0.5 and 1.5 assumed by Grais, de Melo and Urata (1986). The evidence that the aggregate supply elasticity of exports may lie in the range from 1.0 to 4.0 (Goldstein and Khan, 1985: 1,087) also seems to point to higher values for this parameter than those chosen by Grais, de Melo and Urata (1986: 74). The value of 5.0 used in this study was subjectively set somewhat above the empirically estimated values given the well known difficulties involved in obtaining such estimates. Experimentation with higher values for this parameter, and with differential values across commodities, is likely to be of interest.

The elasticities of substitution between exports from China and other export products were set at 10.0 in the belief that Chinese exports of many products are close substitutes for other products in world markets. This assumption is at variance with the few available direct estimates of the elasticity of export demand for China's exports, but the likelihood that such estimates are biased downwards is well known (see, for example, Leamer and Stern, 1970: 56–74). For commodity exports, at least, the value of 10.0 does not seem unreasonable, and is broadly consistent with values used in many other CGE modelling exercises (such as Dixon et al., 1982). Since China's share of world markets remains small, this elasticity of substitution has a greater impact on export demand elasticity than the aggregate world demand elasticity.

The base value elasticity of demand for total world exports was set at -2.0 in light of the relatively low elasticity of substitution between domestic and imported goods assumed in the model. Since the focus of the model is on a relatively short time period, supply adjustment in other countries may be fairly low, placing the major burden of adjustment on demand.

The very high base value for the elasticity of supply of imports to China was chosen to make China essentially a price taker in the market for imports. Given China's small share in most markets (with the notable exception of wool), this assumption, used in most models, does not appear unreasonable.

The elasticity of substitution between primary factors was set to a base value of 0.5. This value was selected by Dixon et al. (1982) after an extensive literature search. While it is substantially below some of the estimates presented in the literature (such as Limskul, 1988), it does not seem unreasonable as a short-run estimate.

As well as uncertainty about the elasticity parameters, there remains some uncertainty about some of the database shares. One area of particular concern, however, is the trade structure implicit in this database. In 1981, for instance, China was a large importer of cotton, whereas, by 1986, it had become a substantial exporter.

Given the potentially very different responses of an exporting or an import-competing industry to changes in the foreign exchange system, this issue was viewed as being particularly serious. Accordingly, the behaviour of the model was examined using the original 1981 trade pattern and an alternative based upon the pattern of trade in 1986.

This adjustment was made in a relatively simple way. The shares of each category of exports (imports) was first adjusted by the change in the share of corresponding category of exports (imports) estimated using data series obtained from the International Economic Data Bank at the Australian National University. The resulting shares were then rescaled to sum to unity, and expressed as shares of total exports (imports) before price adjustment. The share of total exports (imports) was also increased in line with the rise in the share of trade in the Chinese economy over the period (World Bank, 1989b). Finally, the price adjustment factors applied throughout the model were used to adjust the values of exports (imports) in each row of the table. The final balancing of the table was then undertaken using gross output in each sector as the balancing item, as described in Appendix 4A.

Notes

Chapter 1

The authors are grateful for comments from Kym Anderson on an earlier draft. Parts of this chapter originally appeared in Watson, Findlay and Du (1989) and in Findlay and Watson (1990).

1 See Chapter 2 for details.

2 The nature of the system and the effects of reforms are examined in Chapters 3 and 4 by Martin.

3 See Chapters 3 and 4 for details.

4 The composition of these imports is considered in more detail in Chapter 5 by Anderson.

Chapter 2

1 See Maddala (1977) for a discussion of this.

Chapter 3

Particular thanks are due to Christopher Findlay, Peter Drysdale and Kym Anderson for valuable suggestions and advice in the course of this research. Mr D. Thompson, Mr T. Mallawaraachchi and Mr Li Ze provided valuable assistance. I also benefited from comments by participants in seminars at the Australian National University, Adelaide/Flinders Universities, the University of Melbourne, and the Chinese Academy of Social Sciences, as well as the workshops conducted during the course of this project. All responsibility for any remaining errors is my own.

Chapter 4

Particular thanks are due to the volume editor, Christopher Findlay, for his suggestions on the structure of this chapter and also to David Thompson and Thilak Mallawaraachchi for providing excellent research assistance.

Chapter 5

Thanks are due to Christopher Findlay, Ross Garnaut, Sartaj Aziz and Wouter Tims for helpful comments on a longer paper from which this study is drawn and to Prue Phillips of the Australian National University's International Economic Data Bank for providing numerous data printouts.

1 The latter conclusion is still somewhat controversial, however. See, for example, Spraos (1980), Sapsford (1985) and Grilli and Yang (1988).

2 For a more detailed treatment of the above argument, see Anderson (1987; 1989, ch. 2).

3 For further empirical evidence for other Pacific rim countries, see Anderson (1983).

4 The estimate of US$500 is from Perkins (1988: 632). The reason the World Bank suggests China's per capita income is still only US$300 may be simply to ensure China remains eligible for concessional loans that are available only to low-income countries.

5 When the world oil price fell in the mid-1980s China initially expanded the volume of its energy exports to help maintain hard currency earnings, but with the rapidly growing domestic demand for energy that expansion has since slowed down.

6 China's growing imports of wool were almost exactly offset by expanding cotton exports in the mid-1980s but, as pointed out below, the latter is likely to decline before long.

7 For a closer examination of the extent to which government policies have both accentuated and caused fluctuations around the trends described above, see Lardy (1983a), Perkins and Yusuf (1984), Perkins (1988), Sicular (1989), and Anderson (1989, ch. 3) and the references therein.

8 Watson and Findlay note in Chapter 9 in this volume that the wool/sheepmeat producer price ratio fell by two-thirds between 1980 and 1985, during which time mutton output rose by a third while raw wool production grew a mere 1 per cent.

9 Details are provided in Chapter 2 in this volume.

10 Cotton accounted for about 90 per cent of all fibre use in Korea and Taiwan around 1960, as it did in China prior to the mid-1970s. But by the early 1980s only 20–25 per cent of fibre use in Korea and Taiwan was cotton; wool was less than 5 per cent and silk less than 1 per cent. By 1983 China's fibre composition was 80 per cent cotton, 3 per cent wool, 3 per cent silk and 14

per cent synthetic. Given the extremely capital-intensive nature of synthetic fibre production, China is likely to have a strong comparative disadvantage in producing synthetics well into next century (Anderson and Park, 1989).

11 The net addition would be somewhat less because other economies, notably China's East Asian neighbours, would be crowded out of international markets for textiles and clothing by China and hence would import less fibres (Anderson and Park, 1989).

12 For example, Australia, which supplies about half the world's wool exports, substituted out of wool production during the 1970s and back into wool production in the 1980s following changes in the profitability of wool relative to grain and other alternative enterprises. Australia's wool production peaked at 925 kt around 1970, fell to 680 kt in 1978 and was back to a record 950 kt in 1988.

13 For more on this point, see Anderson and Park (1989) and Anderson (1990a). The issues are also reviewed in the introduction to this volume.

Chapter 7

Another version of this study which contained case studies of importers rather than a full set of survey results was presented to the joint workshop of the Economics Department, University of Adelaide and the International Trade Research Institute of the Chinese Ministry of Foreign Economic Relations and Trade, held in Guangzhou (Canton) in January 1988. Comments from participants in that meeting and from Helen Hughes and an anonymous referee are acknowledged. A subsequent version was published in the *Pacific Economic Papers* series of the Australia–Japan Research Centre, Australian National University.

1 Administrative arrangements in world textiles and clothing trade are described by Cline (1987).

2 See various issues of *Textile Asia,* but in particular 'QR in action' by Peter Harding in the June 1989 issue.

3 See Crowley, Findlay and Gibbs (1989) for details of the survey procedure.

4 When country was included as a second treatment, it was significant in all cases but textiles/clothing was not.

5 We also measured the partial correlation coefficients between survey responses and the changes in market shares between 1980 and 1986, but few characteristics were significant and the results are not reported.

6 In an article subtitled 'If the US textile industries cannot gain more protection, technology may help them' (*Textile Asia*, May 1989: 156), the author argues that 'Quick Response ... will give [US industry] a competitive edge over foreign competition'.

7 The official definition of re-exports (by the Census and Statistics Department of Hong Kong) is that it includes processes such as 'diluting, packing, bottling, drying, assembling, sorting and decorating' but excludes such processes which permanently change the share, nature, form or utility of a product.

8 See Yenny and Uy (1985) for an analysis of low truck productivity in China. Louven (1985) discusses the contribution of the strategy of dispersed development in China to lowering returns to investments in the railway infrastructure.

Chapter 8

1 For a further discussion of the agricultural and pastoral districts of China, see Economic Geography Research Section, Geographical Research Institute, Chinese Academy of Sciences (1983). For a more detailed analysis of the ecology and economic geography of the pastoral areas see Zhongguo Xumuye Zonghe Quhua Yanjiuzu [Research Group into the General Regional Divisions of Animal Husbandry in China] (1984). Map 8.2 is drawn from the latter source.

2 The following discussion is based on the *Chinese Agricultural Encyclopedia —Agricultural Meteorology Volume* (1986).

3 For further discussion of the types of natural disasters, see Lin Xiangjin (1982).

4 Examples in Yunnan are cited in Pan Junqian (1987).

5 The theoretical stocking rate is the number of animals that can be sustained by a unit of grassland without causing that grassland to deteriorate.

6 About 33 per cent of the grasslands of Inner Mongolia have deteriorated, and the output of herbage has declined by 30 to 60 per cent compared with the early 1950s. Over the same period, the area of improved and enclosed grassland has amounted to only about 30 million mu. Degeneration is thus much faster than construction. During the Sixth Five-Year Plan (1981–85) the output of grass decreased by 30 to 50 per cent compared with the early 1960s. In Qinghai Province about 20 to 30 million kg of dried herbage is now gathered and stored each year, which is only equal to 25 per cent of that in the 1950s. In Tibet there are about 640 million mu of grassland in use and more than 30 per cent (about 200 million mu) is degraded. In Ningxia about

98 per cent of the grassland has deteriorated. The output of grass had decreased by 35 per cent compared with the 1960s. In Gansu deteriorated prairie and grassland overgrown with poisonous weed make up 30 per cent of the total area. In Sichuan degraded grassland accounts for 30 per cent of the total arable area. For a discussion of these examples, see Ai Yunhang (1987) and Wu Jinghua (1988).

7 For a discussion of China's main breeds, see Editorial Committee of the 'Breeds of Domestic Animals and Poultry in China' (1988).

8 Details are provided in Editorial Committee of the 'Breeds of Domestic Animals and Poultry in China' (1988: 108).

9 Interviews, Xinjiang Academy of Animal Science, November 1989. Similar results were obtained in the Soviet Union (see Maxwell et al., 1988).

10 This figure is usually 60–70 per cent in countries like Australia or New Zealand.

11 A brief summary of these changes can be found in China, State Council Rural Development Research Centre (1987). More detailed descriptions of the precise forms of household contracting and the way they changed can be found in the various regulations published by central, provincial and autonomous region governments. See, for example, the Inner Mongolian reports and regulations for animal husbandry and pasture management in China Agricultural Yearbook Editorial Committee (1985, 1986).

12 The following discussion is based on Liu Delun and Li Zhengqiang (1987) and China, State Council Rural Development Research Centre (1987).

13 The allocation of sheep to households has led to some under-reporting to the State Statistical Bureau and the size of the decline may be slightly overstated in the national figures.

14 The Qinghai Animal Husbandry Bureau Management Division (1988) reports that in Qinghai, labour costs account for over 60 per cent of sheep rearing costs.

Chapter 9

An earlier version of some sections of this chapter appeared in Watson, Findlay and Du (1989).

1 This issue is examined in a more general way in Findlay and Watson (1988).

2 The following discussion is based on Commercial Economics Research Institute, Ministry of Commerce (1984).

3 See Watson (1988) for discussion of the timing of the policy changes.

4 See Wang Bingxiu (1987) for a discussion of this issue.

5 For a discussion of the grain price changes in 1985, see Watson (1989: 116–18).

6 The only other major contributing factor not considered here is the possibility of a major natural disaster affecting the wool-producing areas over this period. While the variable climatic and natural conditions in the producing areas (see Chapter 8) mean that such problems are a regular occurrence (see, for example, the discussion in Lin Xiangjin, 1982), none of the authors cited in this chapter raise this as an explanation for the fluctuations in wool production in recent years.

7 Interviews in Yulin, August 1988.

8 See further the Central Committee and State Council document on 'Ten Policies for Further Enlivening the Rural Economy', 1 January 1985.

9 *Zhongguo Shangye Bao* [China Commercial Paper] 3 July 1986, gives a list of the price increases in different regions after 1984.

10 For a fuller discussion of these issues, see Watson (1989). Useful Chinese summaries of these reforms can be found in Tian Yinyong et al. (1985), Han Bi and Mai Fukang (1984), Xiang Huaicheng (1987) and various issues of *Zhongguo Jingji Nianjian* [China Economic Yearbook].

11 Wang Xukai and Lang Zuoshi (1987) report that in 1985 Suzhou City alone added 10 factories of 5,000 spindles exceeding the total Jiangsu Province Seventh Five-Year Plan target of 15,000 spindles. And in Shanxi Province wool yarn consumption rose by 87.7 per cent during 1984.

12 The version of this article published in *Xinjiang Caijing* [Xinjiang Finance and Economics] No. 6 (1987: 7–11) gives more detail on the methods used.

13 The poor quality of local processing, often using obsolete machinery, is stressed in *Zhongguo Shangye Bao*, 6 October 1988.

14 'It will be hard to solve the major shortage of raw materials in the wool textile industry in the short term' (*Jingji Xiaoxi* [Economic Information], 25 March 1988: 2).

15 Data reported in this section were obtained from interviews in Xinjiang in November 1989, unless otherwise stated.

16 The following discussion is based on reports in *Jingji Cankao* [Economic Reference], 1 February 1990: 4; *Zhongguo Jingji Xinwen* [China Economic News], 19 March 1990: 2; *Nongmin Ribao* [Peasant's Daily], 22 May 1990: 2, and 6 June 1990: 2; and *Zhongguo Shang Bao* [China Commercial Paper], 23 June 1990: 3, and 29 June 1990: 3.

Bibliography

Ai Yunhang (1987) 'Guanyu muqu jingji wenti de yanjiu' [Research on economic issues in pastoral areas] internal publication

—— (1988) 'Wo guo yangmao gong qiu fangmian de zhuyao wenti ji qi duice' [Major issues in wool supply and demand in China and the appropriate policies] *Nongye Jingji Wenti* [Problems of Agricultural Economics] No. 7: 30–4 (July)

Anderson, K. (1983) 'Economic growth, comparative advantage and agricultural trade of Pacific rim countries' *Review of Marketing and Agricultural Economics* Vol. 51, No. 3: 231–48 (December)

—— (1987) 'On why agriculture declines with economic growth' *Agricultural Economics* Vol. 1, No. 3: 195–207 (June)

—— (1988) 'Rent-seeking and price distorting policies in rich and poor countries' *Department of Economics Working Paper* 88–10, University of Adelaide (November)

—— (1989) *Changing Comparative Advantage in China: Effects on Food Feed and Fibre Markets* Paris: OECD Development Centre (forthcoming, also in French)

—— (1990a) 'China and the Multi-fibre Arrangement' in C.B. Hamilton (ed.) *Textile Trade and the Developing Countries: Liberalizing the MFA in the 1990s* Washington DC: World Bank

—— (ed.) (1990b) *New Silk Roads: East Asia and World Textile Markets* Cambridge and New York: Cambridge University Press (forthcoming)

Anderson, K., Y. Hayami and others (1986) *The Political Economy of Agricultural Protection: East Asia in International Perspective* Sydney and London: Allen and Unwin

Anderson, K. and Y. I. Park (1989) 'China and the international relocation of world textile and clothing activity' *Weltwirtschaftliches Archiv* Vol. 125, No. 1: 129–48

Anderson, K. and R. Tyers (1987) 'Economic growth and market liberalisation in China: implications for agricultural trade' *The Developing Economies* Vol. 25, No. 2: 124–51 (June)

Anderson, K. and R. Tyers (1989) *Global Effects of Liberalising Trade in Farm Products* Thames Essay No. 55, Aldershot: Gower, for the Trade Policy Research Centre

Angel, C., P. Simmons and R. Coote (1988) 'Wool consumption in China' *Quarterly Review of the Rural Economy* Vol. 10, No. 1: 70–8

Armington, P. (1969) 'A theory of demand for products distinguished by place of production' *IMF Staff Papers* 16/2: 179–201, IMF

Australian Wool Corporation (1988) 'Wool information paper' AWC: Melbourne (mimeo)

Balassa, B. (1965) 'Trade liberalization and "revealed" comparative advantage' *Manchester School of Economic and Social Studies* Vol. 33, No. 2: 99–124 (May)

—— (1979) 'The changing pattern of comparative advantage in manufactured goods' *Review of Economics and Statistics* Vol. 61, No. 2: 259–66 (May)

Balassa, B. and L. Bauwens (1988) *Changing Trade Patterns in Manufactured Goods: An Econometric Investigation* Amsterdam: North Holland

Ball, K., S. Beare and D. Harris (1989) 'The dynamics of fibre substitution in the raw fibre market: a partial adjustment trans-log model' ABARE paper presented to the Conference of Economists, University of Adelaide, July

Ballou, R. (1978) *Basic Business Logistics* Englewood Cliffs, NJ: Prentice Hall

Beijing Agricultural College Research Group (1987) 'Gaige he wanshan wo guo yang mao liutong tizhi de tantao [An exploration of the reform and improvement of China's system of wool circulation] *Zhongguo Xumuye Fazhan Zhanlue* Yanjiu [Research into the Development Strategy for China's Animal Husbandry Industry] Beijing: Zhongguo Zhanwang Chubanshe

Blejer, M. and G. Szapary (1989) 'The evolving role of fiscal policy in centrally planned economies under reform' *IMF Working Paper* 89/26, IMF

Breusch, T. S. and A. R. Pagan (1979) 'A simple test for heteroskedasticity and random coefficient variation' *Econometrica* 47: 1287–94

Brodin, A. and D. Blades (1986) 'The OECD compatible trade and production data base, 1970–1983' *Department of Economics and Statistics Working Paper* No. 31, OECD Secretariat, Paris (March)

Brummit, W. E. (1989) 'China's economic growth, agricultural self-sufficiency and its livestock and grain sectors' BE (Hons) thesis, University of Adelaide

Byrd, W. (1987) 'The impact of the two-tier, plan/market system in Chinese industry' *Journal of Comparative Economics* Vol. 11, No. 3: 295–308

—— (1989) 'Plan and market in the Chinese economy: a simple general equilibrium model' *Journal of Comparative Economics* 13: 177–294

Byrd, W. and G. Tidrick (1987) 'Factor allocation and enterprisė incentives' in G. Tidrick and J. Chen (eds) *China's Industrial Reform* Oxford University Press, for the World Bank

Carter, C. and F. N. Zhong (1988) *China's Grain Production and Trade: An Economic Analysis* Boulder: Westview Press

Chai, C. H. (1988) 'Economic relations with China' in H.C.Y. Ho and L.C. Chau *The Economic System of Hong Kong* Hong Kong: Asian Research Service

Chan, T. (1989) 'China's foreign exchange and trade controls: what next?' Paper presented at a seminar at the Centre for Chinese Political Economy, Macquarie University, Sydney, August

Chen Xikang (1988) 'The effects of Chinese economic reforms on agricultural and industrial productivity' Institute of Systems Science, Chinese Academy of Sciences, Beijing (mimeo)

Chenery, H., S. Robinson and M. Syrquin (1986) *Industrialization and Growth: A Comparative Study* New York: Oxford University Press, for the World Bank

Chenery, H.B. and M. Syrquin (1975) *Patterns of Development 1950–1970* New York: Oxford University Press

Chey, J. and P. Lau (1989) 'China's demand for wool products' Paper presented at the conference on East Asia and the Redirection of World Trade in Fibres, Textiles and Clothing, Australian National University, August

Chey, S. (1987) 'The Chinese raw wool production industry' Australian National University (November) (mimeo)

China, State Council Rural Development Research Centre (ed.) (1987) *Nongcun Zai Biange Zhong Qianjin: Lai Zi Jiceng De Diaocha Baogao* [The Countryside Advances in the Midst of Change: Survey Reports from the Basic Level] Beijing: Nongye Chubanshe

China, State Planning Commission and State Statistical Bureau (1987) *Input–Output Tables for China* China Statistical Information Centre, Beijing and the East West Center, Hawaii

China, State Statistical Bureau (1983) *China Statistical Yearbook 1983* Hong Kong: Jingji Daobaoshe

—— (1985) *China Statistical Yearbook 1985* Hong Kong: International Centre for the Advancement of Science and Technology

China Agricultural Yearbook Editorial Committee (1985, 1986) *Zhongguo Nongcun Fagui 1983, 1984* [Laws and Regulations for the Chinese Countryside, 1983, 1984] Beijing: Nongye Chubanshe

—— (1988) *China Agricultural Yearbook 1987* Beijing: Agricultural Publishing House

Chinese Agricultural Encyclopedia – Agricultural Meteorology Volume (1986) Beijing: Agricultural Publishing House

Chinese Economic System Reform Research Institute (1987) *Reform in China* Armonk, New York: East Gate Press

Chow, G. (1987) 'Money and price level determination in China' *Journal of Comparative Economics* Vol. 11, No. 3: 319–33

Cline, W.R. (1987) *The Future of World Trade in Textiles and Apparel* Washington DC: Institute for International Economics

Codsi, G. and K. Pearson (1988) 'GEMPACK: general purpose software for applied general equilibrium and other economic modellers' *Computer Science in Economics and Management* 1: 189–207

Commercial Economics Research Institute, Ministry of Commerce (ed.) (1984) *Xin Zhongguo Shangye Shigao* [The History of Commerce in New China] Beijing: Zhongguo Caizheng Jingji Chubanshe

Cross, J. (1989) 'The Chinese wool and textile industries' Paper presented to the Conference of the International Wool Textile Organisation, Perth

Crowley, J., C. Findlay and M. Gibbs (1989) 'China's penetration of the Australian textile and clothing market' *Pacific Economic Papers* No. 175, Australia–Japan Research Centre, Australian National University (September)

Dai Luzhang (1987) 'New steps to deepen the reform of China's banking system' People's Bank of China (mimeo)

Deardorff, A. V. (1984) 'An exposition and exploration of Krueger's trade model' *Canadian Journal of Economics* Vol. 17, No. 4: 731–46 (November)

Deaton, A. S. and J. Muellbauer (1980) *Economics and Consumer Behaviour* Cambridge: Cambridge University Press

de Melo, J. and S. Robinson (1989) 'Product differentiation and the treatment of foreign trade in computable general equilibrium models of small economies' *Journal of International Economics* 27: 47–67

de Wulf, L. and D. Goldsborough (1986) 'The evolving role of monetary policy in China' *IMF Staff Papers* 33: 209–42, IMF

Dervis, K., J. de Melo and S. Robinson (1981) 'A general equilibrium analysis of foreign exchange shortages in a developing country' *Economic Journal* 91: 891–906

Desai, P. and J. Bhagwati (1979) 'Three alternative concepts of foreign exchange difficulties in centrally planned economies' in J. N. Bhagwati (ed.) *International Trade: Selected Readings* Cambridge, Mass.: MIT Press

Diao Xinshen (1987) 'The role of a two tier price system' in Chinese Economic System Reform Research Institute *Reform in China* Armonk, New York: East Gate Press

Dixon, P., B. Parmenter, J. Sutton and D. Vincent (1982) *ORANI: A Multisectoral Model of the Australian Economy* Amsterdam: North Holland

do Rosario, L. (1989) 'One star for debt control' *Far Eastern Economic Review* 19 January: 48–50

Du Yintang (1987) 'The wool market in China' Centre for Asian Studies, University of Adelaide (mimeo)

Eaton, J. (1987) 'A dynamic specific-factors model of international trade' *Review of Economic Studies* Vol. 54, No. 2: 325–38 (April)

Economic Geography Research Section, Geographical Research Institute, Chinese Academy of Sciences (ed.) (1983) *Zhongguo Nongye Dili Zonglun* [A General Outline of China's Agricultural Geography] Beijing: Kexue Chubanshe

Editorial Committee of the 'Breeds of domestic animals and poultry in China' (1988) *Zhongguo Yang Pinzhong Zhi* [Sheep and Goat Breeds in China] Shanghai: Shanghai Scientific and Technical Publishers

Feder, G. (1982) 'On exports and economic growth' *Journal of Development Economics* 12: 59–73

Feltenstein, A. and F. Ziba (1987) 'Fiscal policy, monetary targets, and the price level in a centrally planned economy: an application to the case of China' *Journal of Money, Credit and Banking* Vol. 19, No. 2: 137–56

Findlay, C.C. and A. Watson (1988) 'Efficiency and risk, contracting in the Chinese countryside' Paper presented to the Wool Project Workshop, Australia–Japan Research Centre, Australian National University, 14 October

—— (1990) 'China and Australian wool' *Pacific Economic Papers* No. 180, Australia–Japan Research Centre, Australian National University (February)

Gao Xiaoming (1984) 'Guanyu yangmao shougou zhong de jingji zhengce wenti' [Some policy issues in wool purchasing] *Nongye Jingji Wenti* [Problems of Agricultural Economics] No. 7: 54–5

Garnaut, R. (1988) 'Asia's giant' *Australian Economic Papers* Vol. 27, No. 51: 173–86 (December)

Goldstein, M. and M. Khan (1985) 'Income and price elasticities in international trade' in R. Jones and P. Kenen (eds) *Handbook of International Economics* Amsterdam: Elsevier Science Publications

Grais, W., J. de Melo and S. Urata (1986) 'A general equilibrium estimation of the effect of reduction in tariffs and quantitative restrictions in Turkey in 1978' in T. Srinivasan and J. Whalley (eds) *General Equilibrium Trade Policy Modelling* Cambridge, Mass.: MIT Press

Grilli, E. R. and M. C. Yang (1988) 'Primary commodity prices, manufactured goods and prices, and terms of trade of developing countries: what the long run shows' *World Bank Economic Review* Vol. 2, No. 1: 1–48

Haitovsky, Y. (1972) *Regression Analysis from Grouped Observations* London: Griffin's Statistical Monographs and Courses No. 33

Hamilton, C. (1986) 'Import quotas and voluntary export restraints' in C. Findlay and R. Garnaut (eds) *The Political Economy of Manufacturing Protection: Experiences of ASEAN and Australia* Sydney: Allen and Unwin

Han Bi and Mai Fukang (eds) (1984) *Guojia Yusuan Cankao Ziliao* [Reference Materials on the National Budget] Beijing: Zhongyang Guangbo Dianshi Daxue Chubanshe

Hardin, G. (1968) 'The tragedy of the commons' *Science* 162: 1243–8 (December)

He Juhuang (1989) 'An analysis of Chinese residents' demand' Paper presented to the bilateral Australia–China workshop on The Role of Economic Modelling in National Economic Management, University of Melbourne, 25–29 November

He Long (1990) 'The need to strengthen macro controls over wool' *Zhongguo Shangye Bao* 23 June: 3

Hinchy, M. and B. Fisher (1988) 'Benefits from price stabilization to producers and processors: the Australian buffer-stock scheme for wool' *American Journal of Agricultural Economics* Vol. 70, No. 3: 604–15

Ho Lok Sang (1986) 'Whither China's foreign exchange controls?' Department of Economics, Chinese University of Hong Kong (mimeo)

Indonesia, Biro Pusat Statistik (1984) *Input–Output Tables, Indonesia, 1980* Volumes 1–4, Jakarta: Central Publishing

International Wool Secretariat (IWS) (1987) *Wool Facts* Ilkley: International Wool Secretariat

Isard, P. (1977) 'How far can we push the law of one price?' *American Economic Review* 67: 942–8

Johnson, D. G. (1973) *World Agriculture in Disarray* London: Fontana

Jones, R.W. and H. Kierzkowski (1989) 'The role of services in production and international trade: a theoretical framework' Paper presented to the Second European Workshop on International Trade, Bergen, June

Keesing, D. B. and D. R. Sherk (1971) 'Population density in patterns of trade and development' *American Economic Review* Vol. 61, No. 5: 965–71 (December)

Kis, P., S. Robinson and L. Tyson (1986) 'Computable general equilibrium models for socialist economies' *Division of Agriculture and Natural Resources Working Paper* No. 394, University of California, Berkeley

Kravis, I. B., W. Heston and R. Summers (1983) 'The share of services in economic growth' in F. G. Adams and B. G. Hickman (eds.) *Global Econometrics* Cambridge, Mass.: MIT Press

Kravis, I. B. and R.E. Lipsey (1988) 'National price levels and the price of tradables and nontradables' *American Economic Papers* Vol. 78, No. 2: 474–8 (May)

Krueger, A. O. (1977) *Growth, Distortions and Patterns of Trade Among Many Countries* Princeton, NJ: International Finance Section, Princeton University

Kumpe, T. and P. T. Bolwijn (1988) 'Manufacturing: the new case for vertical integration' *Harvard Business Review* March–April: 75–81

Kuznets, S. S. (1966) *Modern Economic Growth: Rate, Structure and Spread* New Haven: Yale University Press

—— (1971) *Economic Growth of Nations: Total Output and Production Structure* Harvard University Press

Lal, D. (1987) 'The political economy of economic liberalisation' *World Bank Economic Review* Vol. 1, No. 2: 273–300 (January)

Lardy, N. (1983a) *Agriculture in China's Modern Economic Development* Cambridge: Cambridge University Press

—— (1983b) 'Agricultural Prices in China' *World Bank Staff Working Paper* No. 606, Washington DC: World Bank

Lattimore, O. (1962) *Inner Asian Frontiers of China* Boston: Beacon (paperback edition)

Leamer, E. E. (1987) 'Paths of development in the three-factor, good general equilibrium model' *Journal of Political Economy* Vol. 95, No. 5: 961–99 (October)

Leamer, E. and R. Stern (1970) *Quantitative International Economics* Boston: Allyn and Bacon

Li Yuxiang (1986) *Xinjiang Xumu Jingji Gailun* [An Outline of Animal Husbandry Economics in Xinjiang] Urumqi: Xinjiang Renmin Chubanshe

Limskul, K. (1988) 'The sectoral capital stock, employment and sources of economic growth in Thailand: 1960–86' *International Economic Conflict Discussion Paper* No. 40, Economic Research Center, Nagoya University

Lin, C. Z. (1988) 'China's economic reforms 2: Western perspectives' *Asian Pacific Economic Literature* Vol. 2, No. 1: 1–25

Lin (Justin) Yifu (1990) 'The household responsibility system in China's rural reforms' in Alan Maunder and Alberto Valdes (eds) *Agriculture and Government in an Interdependent World* Aldershott: Dartmouth

Lin Shujian and Yang Yongzheng (1987) 'China's exports: performance and issues' Paper prepared for international workshop 'Can World Markets Continue to Sustain Export Orientated Strategies in Developing Countries?', 31 August – 4 September, National Centre for Development Studies, Australian National University, Canberra

Lin Xiangjin (1982) *Wuoguo Xumuye Jingji* [China's Animal Husbandry Industry] Beijing: Nongye Chubanshe

Liu Delun and Li Zhengqiang (1987) 'Mian yang shengchan guanli wenti tantao' [A study of the problems in the management of sheep production] *Zhongguo Nongcun Jingji* [Chinese Rural Economy] 5 (May)

Liu Deyou (1986a) 'Yangmao dazhan xu pingxi' [The 'wool war' urgently needs calming down! *Jingji Ribao* [Economics Daily] 15 November: 1, 2

—— (1986b) 'Yangmao dazhan ru he pingxi' [How to calm the 'wool war'?] *Jingji Ribao* [Economics Daily] 17 November: 2

—— (1988) 'Yangrong da luan ru he pingxi' [Can the 'cashmere chaos' be cured?] *Jingji Ribao* [Economics Daily] 4 January: 2

Lluch, C., A. A. Powell and R. A. Williams (1977) *Patterns in Household Demand and Savings* New York: Oxford University Press

Louven, E. (1985) 'Development and distribution in China' Department of Economics Seminar Paper, University of Hong Kong (mimeo)

Lu Xuezeng, Yang Shengming and He Juhuang (1985) 'The structure of China's domestic consumption, analysis and preliminary forecasts' *World Bank Staff Working Paper* No. 755, World Bank

Maddala, G. S. (1977) *Econometrics* New York: McGraw-Hill

Martin, W. (1990a) 'Implications of China's foreign exchange system for the wool market' *Pacific Economic Papers* (forthcoming) Australia–Japan Research Centre, Australian National University, Canberra

—— (1990b) 'Modelling the post-reform Chinese economy' *China Working Paper* 90/1, National Centre for Development Studies, Australian National University, Canberra

—— (1990c) 'Two-tier pricing in China's foreign exchange market' *National Centre for Development Studies Working Paper* (forthcoming) Australian National University, Canberra

Martin, W. and S. Suphachalasai (1990) 'Effects of the Multi-Fibre Arrangement on developing country exporters: a simple theoretical framework' in C. Hamilton (ed.) *Textiles and the Multi-Fibre Arrangement* Washington: World Bank (forthcoming)

Maxwell, W. M. C., S. Salamon, I. W. Purvis and M. A. O'Flynn (1988) 'Australia–USSR Agricultural Cooperation Agreement' Report on the Australian Mission to the USSR on Merino Sheep Breeding, 23 September to 6 October, Department of Primary Industries and Energy, Canberra

McMillan J., J. Whalley and L. Zhu (1987) 'The impacts of China's economic reform on agricultural productivity growth' *Journal of Political Economy* Vol. 97, No. 4: 781–807

Men Xiuqi (1990) 'The current situation in the wool textile industry and the way out' *Zhongguo Xiangzhen Qiye* 7 March: 2

Mohamed Yusupu (1987) 'Developing the pastoral economy: selected works from the national conference on the pastoral economy' internal publication

Naughton, B. (1985) 'False starts and second wind: financial reforms in China's industrial system' in E. Perry and C. Wong (eds) *The Political Economy of Reform in Post-Mao China* Harvard Contemporary China Series 2, Harvard University Press

Niu, R. and P. Calkins (1986) 'Towards an agricultural economy for China in a new age: progress, problems, response and prospects' *American Journal of Agricultural Economics* Vol. 68, No. 2: 445–50

Pan Junqian (1987) 'Wuoguo mianyang shengchan de heli jiegou' [The rational structure of sheep production in China] internal publication

Park, Y. I. and K. Anderson (1989) 'The rise and demise of textiles and clothing in economic development: the case of Japan' Centre for International Economic Studies Seminar Paper 89–06, University of Adelaide (August)

Perkins, D. H. (1986) *China: Asia's Next Economic Giant?* Seattle: University of Washington Press

—— (1988) 'Reforming China's economic system' *Journal of Economic Literature* Vol. 26, No. 2: 601–45 (June)

Perkins, D. H. and M. Syrquin (1988) 'Large countries: the influence of size' in H. Chenery and T. N. Srinivasan (eds) *Handbook of Development Economics* Vol. 2, Amsterdam: North Holland

Perkins, D. H. and S. Yusef (1984) *Rural Development in China* Baltimore: John Hopkins University Press

Powell, A. A. and F. Gruen (1968) 'The constant elasticity of transformation production frontier and the linear supply system' *International Economic Review* Vol. 9, No. 3: 315–28

Prais, G. J. and J. Aitchison (1954) 'The grouping of observations in regression analysis' *Review of International Statistical Institute* 22: 1–22

Qian Fenyong (1987) 'Jiasu muqu jingji fazhang de jianyi' [Proposal to promote economic development in pastoral areas] internal publication

Qinghai Animal Husbandry Bureau Management Division (1988) 'Qinghai caoyuan xumuye jingji xiaoyi qian tan' [A broad discussion of economic results in pastureland animal husbandry in Qinghai] *Nongcun Caiwu Kuaiji* [Rural Financial Accounting] No. 4

Robinson, S. (1989) 'Multisectoral models of developing countries: a survey' in H. Chenery and T. N. Srinivasan (eds) *Handbook of Development Economics*, Vol. 2, Amsterdam: North Holland

Roemer, M. (1986) 'Simple analytics of segmented markets: what case for liberalisation?' *World Development* Vol. 24, No. 3: 429–39

Sapsford, D. (1985) 'The statistical debate on the net barter terms of trade between primary commodities and manufactures: a comment and some additional evidence' *Economic Journal* Vol. 95, No. 379: 781–88 (September)

Schultz, T. W. (1945) *Agriculture in an Unstable Economy* New York: McGraw–Hill

Science Publishing House (1977) 'Neimenggu zizhiqu dongxibu bilin diqu qihou yu nongmuye de guanxi' [The relationship between climate and farming in inner Mongolia and its contiguous zone] Beijing: Science Publishing House

Shan, W. (1989) 'Reforms of China's foreign trade system' *China Economic Review* Vol. 1, No. 1: 33–55

Shirk, S. L. (1985) 'The politics of industrial reform' in E. Party and C. Wong (eds) *The Political Economy of Reform in Post-Mao China* Harvard Contemporary China Series 2, Harvard University Press

—— (1988) 'Plan and market in China's agricultural commerce' *Journal of Political Economy* Vol. 96, No. 2: 383–7

Sicular, T. (1985a) 'Rural marketing and exchange in the wake of recent reforms' in E. Perry and C. Wong (eds) *The Political Economy of Reform in Post-Mao China* Harvard Contemporary China Series 2, Harvard University Press

—— (1985b) 'China's grain and meat economy: recent developments and implications for trade' *American Journal of Agricultural Economics* Vol. 67, No. 5: 1055–62 (December)

—— (1986) 'Prospects and some policy problems of agricultural development in China' *American Journal of Agricultural Economics* Vol. 68, No. 2: 458–60

—— (1987) 'Food price policy in China' *Development Discussion Paper* No. 252, Harvard Institute for International Development

—— (1988a) 'Plan and market in China's agricultural commerce' *Journal of Political Economy* Vol. 96, No. 2: 283–307 (April)

—— (1988b) 'Agricultural planning and pricing in the post-Mao period' *The China Quarterly* 116: 671–705 (December)

—— (1989) 'China: food pricing policy under socialism' in T. Sicular (ed.) *Food Pricing Policy in Asia* Ithaca: Cornell University Press

Simmons, P., B. Trendle and K. Brewer (1988) 'The future of Chinese wool production' ABARE paper presented to the 32nd Annual Conference of the Australian Agricultural Economics Society, 8–12 February, La Trobe University, Melbourne

Siriwardana, M. (1986) 'An initial model of the Chinese economy: a framework for discussion' Bureau of Agricultural Economics Seminar Paper (June), Canberra (mimeo)

Song Zhaoqing et al. (1988) *Songnen Caoyuan Shengtai Xumuye Liangxing Kunhuan Chutan* [A Preliminary Discussion of a Beneficial Cycle of Ecological Animal Husbandry in the Song–Nen Grasslands] Beijing: City Economic Publishing House

Spraos, J. (1980) 'The statistical debate on the net barter terms of trade between primary commodities and manufactures' *Economic Journal* Vol. 90, No. 357: 107–28 (March)

Stoeckel, A. (1979) 'Some general equilibrium effects of mining growth on the economy' *Australian Journal of Agricultural Economics* Vol. 23, No. 1: 1–22

Summers, R. and A. Heston (1988) 'A new set of international comparisons of real product and prices: estimates for 130 countries, 1950–85' *Review of Income and Wealth* (March)

Sung Yunwing (1987) 'The role of microeconomic reforms in the decentralization of China's foreign trade' *China Working Paper* 87/1, National Centre for Development Studies, Australian National University, Canberra

—— (1988) 'A theoretical and empirical analysis of entrepot trade: Hong Kong, Singapore and their role in China's trade' in Leslie V. Castle and Christopher Findlay (eds) *Pacific Trade in Services* Sydney: Allen and Unwin

Sung Yunwing and T. M. Chan (1987) 'China's economic reforms: the debates in China' *Asia Pacific Economic Literature* Vol. 1, No. 1: 1–24

Tam Onkit (1987) 'Issues in foreign exchange control in China', Australian Defence Force Academy, Canberra (mimeo)

Tarr, D. (1989) *A General Equilibrium Analysis of the Welfare and Employment Effects of US Quotas in Textiles, Autos and Steel* Washington: Bureau of Economics Staff Report to the Federal Trade Commission

Tian Yinyong, Zhu Fulin and Kiang Huaicheng (1985) *Lun Zhongguo Caizhen Guanli Tizhi De Gaige* [On the Reform of China's System of Financial Administration] Beijing: Jingji Kexue Chubanshe

Theil, H. and K. W. Clements (1987) *Applied Demand Analysis: Result from System-Wide Approaches* Cambridge: Ballinger

Trela, I. and J. Whalley (1988) 'Do developing countries lose from the MFA?' *National Bureau of Economic Research Working Paper* No. 2618, Cambridge, Mass.: National Bureau of Economic Research

Tyers, R. and K. Anderson (1986) 'Distortions in world food markets: a quantitative assessment' *World Bank Background Paper* No. 22, prepared for the World Bank's World Development Report 1986 (mimeo), Washington DC (January)

—— (forthcoming) *Distortions in World Food Markets* Cambridge: Cambridge University Press

US Department of Agriculture (1988) 'Estimates of producer and consumer subsidy equivalents: government intervention in agriculture, 1982–86 *ERS Staff Report* No. AGES 880127, Washington DC: US Department of Agriculture

—— (1990) *Cotton: Situation and Outlook* Washington DC: US Department of Agriculture

Van der Gaag, J. (1984) 'Household consumption in China: a study of people's livelihood' *World Bank Staff Working Paper* No. 701, Washington DC: World Bank

Vincent, D. P. (1985) 'Exchange rate devaluation, monetary policy and wages: a general equilibrium analysis for Chile' *Economic Modelling* Vol. 2, No. 1: 17–32

Wang Bingxiu (1987) 'Wuoguo yangmao gongqiu zhuangkuang ji qi duice de tantao' [Wool supply and demand and the appropriate policies for China] *Nongye Jingji Wenti* [Problems of Agricultural Economics] No. 3: 11–14

Wang Xukai and Lang Zuoshi (1987) 'Maofangye de weiji yu yangyangye de kunjin ji chulu' [The crisis in the wool spinning industry, the problems of sheep rearing and the solution] *Nongye Jingji Wenti* [Problems of Agricultural Economics] No. 3: 15–17

Watson, A. (1987) 'The family farm, land use and accumulation in agriculture' *Australian Journal of Chinese Affairs* 17: 1–27 (January)

—— (1988)'The reform of agricultural marketing in China since 1978' *The China Quarterly* No. 113: 1–28 (March)

—— (1989) 'Investment issues in the Chinese countryside' *The Australian Journal of Chinese Affairs* No. 22: 85–126 (July)

Watson, A., C. Findlay and Du Yintang (1989) 'Who won the wool war?' *The China Quarterly* No. 118: 213–41 (June)

Whalley, J. (1990) 'The Multifibre Arrangement and China's growth prospects' in K. Anderson (ed.) *New Silk Roads: East Asia and World Textile Markets* Cambridge and New York: Cambridge University Press

Wong, C. (1985) 'Material allocation and decentralisation: impact of the local sector on industrial reform' in E. Perry and C. Wong (eds) *The Political Economy of Reform in Post-Mao China* Harvard Contemporary China Series 2, Harvard University Press

World Bank (1985a) *China: Economic Structure in International Perspective, Annex 5 to China: Long-Term Development Issues and Options* Washington DC: World Bank

—— (1985b) *China Economic Model and Projections, Annex 4 to China: Long-Term Development Issues and Options* Washington DC: World Bank

—— (1988a) *Price Prospects for Major Primary Commodities* (in three volumes) World Bank: Washington DC (November)

—— (1988b) *China: External Trade and Capital* Washington DC: World Bank

—— (1989a) *World Development Report 1989* Washington DC: World Bank

—— (1989b) *World Tables* Washington DC: World Bank

Wu Jinghua (1988) *Dangqian Xumuye Jingji De Rougang Wenti Jiqi Duice* [Some Current Problems in Animal Husbandry Economics and the Appropriate Policies] Beijing: City Economy Publishing House

Wu Jinglian and Zhao Renwei (1987) 'The dual pricing system in China's industry' *Journal of Comparative Economics* Vol. 11, No. 3: 309–18

Xiang Huaicheng (1987) 'Zai gaige zhong qianjin de zhonguo caizheng' [China's financial administration advancing through reform] *Caizheng Yanjiu* [Financial Research] No. 2: 1–9

Yang Yongzheng and R. Tyers (1989) 'The economic costs of food self-sufficiency in China' *World Development* Vol. 17, No. 2: 237–53

Yenny, J. and L.V. Uy (1985) *Transport in China* Washington DC: World Bank

Zhang, Z. (1987) 'Reforming China's exchange rate system: basic problems and options for the future' Shanghai: East China Normal University (mimeo)

Zhang Shaojie and Zhang Amei (1987) 'The present management environment in China's industrial management' in Chinese Economic Reform Research Institute *Reform in China* Armonk, New York: East Gate Press

Zhao Zekun (1986) 'Yangmao shichang: qi shi lu' [The wool market: a revealing story] *Nongmin Ribao* [Peasants' Daily] 29 December: 2

Zheng Peiliu (1985) *Zhongguo Jiaxu Pinzhong Jiqi Shengtai Tezheng* [Animal Varieties and their Ecological Characteristics] Beijing: Agricultural Publishing House

Zhongguo Xumuye Zonghe Quhua Yanjiuzu [Research Group into the General Regional Divisions of Animal Husbandry in China] (1984) *Zhongguo Xumuye Zonghe Quhua* [China's Animal Husbandry Regions] Beijing: Nongye Chubanshe

Zhou Guohua, Qu Weiying and Feng Cheng (1986) 'Xibei "yangmao dazhan" tichu de xin keti' [New issues raised by the 'wool war' in the north-west] *Liaowang* [Outlook] No. 8: 31

Zhou, Z., J. Dillon and G. Wan (1988) 'Development of township enterprise in China' Paper presented at the Annual Conference of the Australian Agricultural Economics Society, La Trobe University, Melbourne, 8–11 February

Index